THE SCHOOL AND COMMUNITY RELATIONS

THE SCHOOL AND COMMUNITY RELATIONS

By

Dr. D.B. Rao
M.Sc., M.A., M.A., M.Ed., Ph.D. (Education)
Reader & Research Director
R.V.R. College of Education
Guntur – 52 006 (A.P.)
(India)

D P H

DISCOVERY PUBLISHING HOUSE PVT. LTD.
NEW DELHI-110002

First Published-2009

ISBN: 978-81-8356-434-2

Published by:

DISCOVERY PUBLISHING HOUSE PVT. LTD.
4831/24, Ansari Road, Prahlad Street
Darya Ganj, New Delhi-110 002 (India)
E-mail: dphtemp@indiatimes.com
dphbooks@rediffmail.com
Website: www.discoverypublishinghouse.com

Printed at:
Arora Enterprises
Laxmi Nagar, Delhi–110 092

Contents

Preface *vii*

1. **Community Relationships** **1**

Basic Issues • Various Approaches • Motivating Areas • Education Commission's Role • Educational Movement • Doorstep Education • Role of Universities • Government and Education • Fast Progress • Targets and Aims • Different Streams • Values Applied • Subject's Importance • Planning Regularly

2. **Community Factors** **37**

Social Elements • Social Effects • The Standings • The Classification • Aims and Factors • Interconnections • Division of Society

3. **School and Community** **59**

School: a Community Centre • Community Serving Methods • Amendment Projects • The Values • Parent-Teacher Collaboration • Significance of Parents' Day • The Curriculum • The Requirement • The Problems • Securing Cooperation Methods • Glorified Essence

4. **Significance of Society** **71**

Necessary Provisions • Political Features • Social Categories • Different Social Systems

5. Framework of Society 127

Modern Line • Many a Change • Change in Society • Process in Continuation

6. Social Empowerment 149

Chance for Equality • The Legacy • Recent Development and Approaches • Related Issues • Various Problems • Employed Women • The Physically Handicapped • The Bizarres • Sufferer Children • Practical Standards • Future Strategies • Fundamental Variations • Historical Ideas • Study Area • NGO's Role • Guiding Theories • General Aspects and Approaches • Ethnic Factors • New Society • Regular Struggle • Gandhi's Role • Political Disinformation • Governments' Role

7. Social Justice by Education 219

Social Variations • Compulsory Education • Purposes and Goals • Fundamental Difficulties • Social Features • Part of Democracy

8. Education for Liberty 239

System Improving • Process of Planning • Process of Urbanisation • Effect of Society

9. Womens' Education 269

The Perceptiveness • Decisive Conditions • Contradicting Approach

10. Democratic Aspects 281

Status of Education • Different Dimensions • The Basics • Social Concerns

Bibliography 289

Index 297

Preface

People, by and large, know the worth of education. They strongly feel that they can not make progress unless and until they are educated. In the very beginning of civilization, men in society had started taking interest in education and after some time, when considerable number of men became educated, they made extensive efforts to educate their women also. The story is same, the world over.

In India, although education has on age old tradition, yet the modern education prospered in 19th century, under the sway of British regime only. In later years, when people became aware of their importance by virtue of education, the country felt and experienced a great change and social upheaval. Then, in a few decades, revolutionary changes took place, the effects of which can be found even today, rather, what we see today is, but the fruition of the efforts made by the pioneers.

Books of social history are deplete with valuable information about the role of school in social upliftment and the community renaissance. Present modest work is but an humble attempt in the same direction, with some characteristics of its own.

Hopefully, it would be accorded a warm reception from all concerned quarters. For further enhancement, enlightened feedback of the readers is heartily welcome.

Community Relationships

Basic Issues

The National Education Commission has opined that education does not end in the school. It is a life long process. The adult today seriously needs to understand the rapidly changing world and the increasing complexities of social life. Even those who have obtained considerable education should start learning once again because its alternative is isolation from the present reality.

The objective of reconstructing our nation can be fulfilled only when each individual continues to receive the latest education so that he may use his intellect skilfully and determinedly in order to bring about economic development, social change and social security. The Education Commission has rightly pointed out that the farmer who turns over the soil or the worker who manipulates machines must understand the nature of the soil or the machine. He must be aware of the scientific process inherent in productive activity so that he may be able to make full use of new achievements and bring about progress.

Maulana Abul Kalam Azad expressed his views on the importance and function of social education by saying that social education means total human education. It grants the individual literacy so that the knowledge of the world may be available to him. It teaches him how to adapt himself to his environment. This implies giving him education in advanced skills and modes of production so that he may achieve the highest economic level. Its purpose is also to give knowledge of the basic principles of hygiene so that our domestic life may be healthy and prosperous.

This will benefit the individual as well as society. Through this education, we should impart to him the lesson of citizenship so that he may obtain some knowledge of the world and also be enabled to participate in those decisions of the government which are conducive to peace and progress. However, the path to adult and social education is littered with many concomitant problems, that is, those problems which always raise obstacles in the path of social education. These problems are described in the following paragraphs.

Vague Policy on Adult Education: Ever since the achievement of independence, our policy on adult education and the programmes formulated for it have invariably been vague and undefined. An effort was made to achieve clarity in the community development programme, but even that was partial or biased, since it was intended only for rural adults. It has no provisions for adults in urban areas.

Lack of Finance: The finance so far invested in adult education has proved inadequate, with the result that this programme, which should have been given the highest priority in national development, fell behind. At least Rs 27 crores should be spent annually upon this programme.

Absence of a Comprehensive Plan: This programme was accepted in principle, but it was never implemented in a suitably comprehensive manner. Had this been done, the problem would have been over come a long time ago. Dr. Frank went as far as to say categorically that if one literate Indian makes another Indian literate, the problem will be overcome within five years.

Problem of Curriculum: The curriculum of adult education has itself proved a problem, because adults were imparted a certain amount of literacy, but no knowledge of anything else, though this can be done. The method adopted for adults is the same as the one used for teaching children, though the needs, attitudes and abilities of adults are completely different. The curricula designed for them must necessarily take into account their particular needs.

Methods of Teaching: The curriculum should fulfil the needs of adults while the methods of teaching should be such that they help in the achievement of this goal. For the spread of literacy, sentence-charts, diagrams, stories, simple words, radio, television, libraries and the drama stage, etc., can be used successfully. But, in reality, because of a paucity of resources, the techniques of adult education are not brought into use in our country to any noticeable extent.

Problem of Teachers: The problem of teachers for adult education has existed from the very beginning. Under these circumstances, an experiment was made by using teachers of primary schools for this purpose, but it proved a failure because the additional allowance given to the teachers was too nominal. In addition, the extension officers, social educators and rural service workers exhibited complete indifference.

Problem of Continuing Education: Continuing education obviously refers to an educational process which always goes on, and in the sphere of adult education, it refers to the education imparted after literacy has been achieved. If facilities for such education are available after literacy has been imparted, some benefit can be derived from it.

The Agencies of Social Education: Apart from the official agencies involved in social education, some voluntary agencies, are also usually, engaged in this programme, but in India, there is a dearth of such agencies. However, the Literacy House in Lucknow and the Adult Education Council in Delhi are doing this work. These institutions aim at producing trained social educators. Only when there is a growth in the number of such agencies can we hope that social education will make rapid progress.

Administration: Social education has a very vast field and there is no uniformity in administration. The various ministries of the government of India have made efforts to train and educate their employees. The Central Education Advisory Committee has organised conferences and seminars from time to time.

Social Education Institutions: Institutions engaged in social education do not get help at the right time and consequently their enthusiasm dies out prematurely.

Training of Workers: There has also been, in general, a dearth of trained workers in the sphere of social education, because of which the Education Advisory Committee has recommended that such workers be trained.

Post Social Education: It is necessary to make arrangements for further education of adults after making them literate.

The Education Ministry of the government of India has suggested the following measures for overcoming the difficulties:-

1. For individuals of 13 or 14 years of age, primary schools should arrange classes of two hours each at least three or four times a week. This work should be apart from the regular teaching programme of these schools.
2. Teachers should be trained for imparting education to the 14-15 age group. The following programmes should be taken up simultaneously:-
    a. Regular classes should be arranged for men and women over a five month period.
    b. The Neighbouring Group Method should be adopted, under which the teacher should educate groups of people in his neighbourhood. Experience has shown that adults show laziness and unwillingness in coming to teaching centres, and hence, it may be more practicable for the teacher to go to their centres instead.
    c. A Home Class movement must be initiated for the education of women, which can be made an integral part of the Rural Education Programme. Students and teachers should offer their cooperation.
    d. Teacher training establishments should formulate and implement programmes of adult education in their respective communities.
    e. Voluntary organisation should organise adult education schemes.

In the opinion of the Indian Adult Education Committee, the important work of literacy must not be lost sight of in the broader perspective of social education. Literacy should remain an important element in social education because, through the written and spoken work, it affords the adult a powerful medium for becoming educated.

What is needed today is that the various programmes of social education must be organised and harmonised. The objective, should be clear, and the adults must be strongly motivated to receive social education. Success in this endeavour depends upon local leadership and local agencies.

Gandhiji said that illiteracy was a blot upon India's fair name. This blot must be wiped out with all expediency. The objective of adult education is not merely literacy, it must be linked with the daily life of adults. From this viewpoint, the literacy programme must have three elements: i) it should be based on work so that it may impart skills to the learner, ii) illiterate individuals should be able to think about the nation's problem, and iii) adequate skills in reading, writing and mathematics should be developed so that informal education can then be continued indefinitely.

The Four Point Programme for social education is as follows:

Literacy Classes: By 1953, 400,000 individuals were made literate through 40,000 classes. By 1955, 7,000 classes provided literacy to 89,000 individuals. In community development centres, 85,000 classes helped to make 600,000 individuals literate.

Community Centres: Community centres perform the functions of providing literacy, recreation and information.

Youth Clubs: Youth clubs have been organised in villages with the result that village protection groups, youth welfare centres and other such units work for social education have come into existence.

Women's Committees: In each development centre, women's welfare groups and women's councils should be set up so that women can be given information about and involved in hymn singing, social discussion groups, use of smokeless stoves, knitting, stitching, nutrition and balanced diet, domestic skills, general health care, domestic economics, home decoration, literacy, etc.

The following means or instruments are employed in social education: a) the common audio-visual aids, b) black-boards and chalks, c) charts and flash cards, d) flannel and other materials, e) other audio-visual materials, f) films, g) radio, and h) television.

The Education Commission has expressed the views that every possible effort should be made to overcome these obstacles, even though 20 years may pass in the process. It has suggested that: 1) for the 6-11 year age group, universal education must be achieved in 5 years, 2) arrangements must be made for part time education, and 3) arrangements must be made for vocational education.

Various Approaches

The Commission has suggested that the following two methods should be adopted for this purpose:

Selective Approach: In this context, those groups are included which can be controlled easily, it also includes vocational training, in which industrial and commercial enterprise, large industrial enterprises and the government itself can take up budgets against illiteracy. In this method, the following fields can be comprehended:

1. Most workers in industrial establishments are illiterate, and hence such establishments can be required to make arrangements for the education of their own employees.
2. The state should make arrangements for the education of their own employees.
3. The state should make arrangements for the education of all illiterate classes concerned with social and economic growth, since the number of such individuals is very large.
4. Programmes for imparting literacy should be an invariable part of every scheme of social education.

Mass Approach: This is concerned with developing a general method of public or mass education in which advantage will be taken of all educated men and women for eradicating the illiteracy of their brothers and sisters. Voluntary organisations can take effective steps in this direction. Existing schools can be converted into community centres. Regarding this approach, the Kothari Commission has argued that a movement for collective or community literacy depends [illegible] upon the voluntary services of all educated individuals, including governmental as well as non-governmental employees and other sections of society. It has suggested that students of the primary, secondary, undergraduate students and students of vocational schools should be engaged in the teaching of adults as a part of the National Service Scheme.

This Commission has made the following suggestions for accelerating the literacy movement:

1. Organising a curriculum for adult women through the central social welfare board, with 'rural sisters' teaching women in the rural areas with the help of the community.
2. Generating social skills.
3. Adopting education as a dynamic process.
4. Running *ad hoc* courses.
5. Training through units which should be for teachers and common people. In this, students should be given suitable opportunities, and use should also be made of radio and television.
6. Generating a suitable atmosphere before initiating the programme.
7. Preparing the adults psychologically to make them receptive.
8. Making use of such means as radio, television, films, lectures, etc., so as to make teaching suitable.
9. Making schemes for literacy in accord with local circumstances.
10. Making the programmes attractive.
11. Seeking the full cooperation of universities, industries, etc.
12. Organising an effective administration.

Motivating Areas

The field of social education presents vast opportunities for educating the public. Various kinds of efforts have been made for educating people all over the world, and for this purpose various experiments were conducted. The USA and USSR despite the fundamental differences in their political and economic structure, both accept and understand the importance of kind of education.

There, the schools, colleges, clubs and institutes and other institutions which work for improving the standards of the fore mentioned have been provided with extensive cultural facilities. In India, too, various experiments have been conducted in the sphere of social education. These are:

Organisation of Tours and Camps: In such schemes, adults are sent to camps for completing a predetermined course of study which aims at teaching them reading, writing and healthy citizenship. K.G. Saiyadain has laid special emphasis upon the utility of such organisations, because in camps, the atmosphere generated is suitable for the growth of brotherhood as well as leadership. However, it must also be kept in mind that the organisers of tours and camps should have adequate knowledge of mass psychology. These camps can prove to be the nerve-centres of social education, and so they should be managed with sympathy and understanding.

Janata (Public) Colleges: Like the 'folk schools' of Denmark, Janata or public colleges were established in India, too. After observing their work, one observer commented that the most important element was not the quantum of knowledge obtained by the students but the fact that their thoughts and sentiments were stimulated. It is conceivable that they would forget much of what they had learnt, after they left school, but after leaving school, they appear to be completely changed. By that time, they have learnt to listen, óbserve, think and make the best possible use of their abilities.

According to Pyare Lal Rawat, the foundation stone for a public and cooperative life will be laid in the social centres in rural society. Janata colleges have been set up to produce educated and trained workers for such social centres. Leadership will take birth in these colleges. They will not have the form and structure of the ordinary colleges because their purpose is to train young men and women to perform activities of social, cultural and educational value among the rural people so that they may be able to provide leadership to the rural areas. In other words, their purpose is to generate knowledge and education among the rural people, to teach the rural people better ways of living so that they may lead a better and more contented life. The objective of these colleges will be to make Indian rural people responsible citizens of a democratic society by arousing their civic, social and cultural awareness.

Community Centres: Community centres were set up in rural areas to provide education, sociability, recreation as well as education in good citizenship. These centres were set up in the village squares (chaupals), panchayat houses and schools, and provided with such materials as carpets, furniture, books, radios, newspapers, periodicals, sports goods, etc.

Intensive Libraries: Where community centres have not been established, intensive libraries have been set up. These libraries contain literature suitable for the adults, as well as books on agriculture, cottage industries, cooperative societies, health and other such subjects of utility.

Educational Caravans: Educational caravans were put into use in the rural areas around Delhi. These caravans are equipped with film projectors, health materials, libraries and other teaching materials. This caravan sets up camp at one place, and then its workers spread all round, making arrangements for adult education.

Adult Education Departments in Universities: The Rajasthan university has made adequate efforts towards adult education by setting up a department, which has also initiated a number of schemes for continuing education. Other universities should imitate Rajasthan in this respect. Universities can make their contribution to social education through postal courses, short-term education, extension services, post-education contact programmes, etc.

Mass Education Movement: 20 years ago, the 'Parichaya' (acquaintance) trust was set up for the spread of mass education for the Gujarati speaking people. So far, this trust has published as many as three hundred books on various subjects. This movement was strengthened by the activities of Wadilal Dagli (Editor, *Commerce*) and Yashwant Doshi (Editor, 'Parichaya' trust publications). Dagli was inspired in this endeavour by the Chicago University pamphlets; which he saw at Berkeley (California). He thought of spreading mass education in his mother tongue. The work began with only Rs 500. Each pamphlet was priced at 0.40 paise. In the first year, the organisation enrolled six hundred members. The publication was based on the principle of giving 10 per cent royalty to authors and Commission to the middle men.

This movement brought about a social revolution, and also provided the illiterate people with a ray of hope. So far, the trust has published 300 pamphlets, thus providing the neo-literates with reading material on a variety of subjects. The trust now plans to publish similar literature for neo-literates in other languages also.

Informal Education: In the fifth plan, one important experiment in the sphere of social education was that of informal education. In this context, the Central Education Advisory Council has prepared a scheme for linking social education with national development. The scheme envisages providing further opportunities for education to those who were compelled to abandon their education too early, Correspondence courses, evening classes, etc.; are part of this scheme. It will provide free and informal opportunities of education for everyone.

Village Education Muhim: This experiment was carried out in Maharashtra. Some social workers conducted adult education classes. In. Vidharbha, some enthusiastic workers were given training and the number of adult education centres was thus increased. This movement was based on the declaration of the workers that they would maintain their own literacy, send their children regularly to school, maintain unity in the village and seek to fulfil the goal of integral development of the village.

Now, this programme is being conducted by District Councils. It has three objectives:

1. Putting an end to illiteracy among adults (14-50 age group)
2. Granting neo-literates opportunities to remain in constant contact with education.
3. Bringing about the comprehensive development of villages through social education centres.

Village Vidyapeeth: This experiment was initiated in Mysore. Through it, the spread of education in economics, agricultural education, etc., was achieved. The project received inspiration from the Danish folk high schools. In these institutions, a class has to remain at the school for 5 or 6 months for studies. This helps to develop a harmonious community life.

Punjab Literacy Project: Mrs. Helen Butt developed an intensive literacy method (Hindi) in the Neelokheri block of Karnal. After its initiation by Mrs. Butt, the project continued with the help of 20 teachers, 7 lady teachers (planning and panchayat department), and 4 supervisors. Its success is proved by the fact that 36 classes for men and 11 classes for women were started. 1000 adults attended these classes. Each full time teacher taught two classes on alternate days. Classes were held on three days a week basis over a period of six months. The learners were taught to read and write Hindi. In addition to literacy, a programme of continuing education was also conducted over 3 months. A short nine-day programme for training teachers was also conducted. Besides, literacy was spread through the Immediate Learning Method. When the project was introduced at the state level, the farmers showed exceptional enthusiasm for it, proved by the fact that attendance in the classes was as much as 19.4 per cent. Mrs. Helen Butt has said after her experience of this project that the total expenditure on adult literacy scheme should be borne by the education department and the institutions (panchayats) which stand to benefit from them.

Adult Education Department Rajasthan University Jaipur: In 1965, the Rajasthan University at Jaipur set up a department of adult education in order to explore the possibilities of adult education by spreading continuing education. This department organises conferences and seminars at regular intervals. It also organises a variety of short-term courses for the neo-literates as well as-the semi-literates. This department has done the following kinds of work with the assistance of the government, the university and voluntary organisations: i) studies in the philosophy and concept of adult education, ii) education in literacy- social, economic development and literacy, adult literacy and the methods of teaching, teacher training, village education muhim, iii) portable audio visual materials, library, conferences, discussions, iv) liberal education, v) public cooperation, vi) training in leadership (Mysore

Vidyapeeth), vii) the sphere of continuing education industrial and vocational development, viii) education of adult women, and ix) agencies and organisation. This department continues to think upon and discuss such issues as the university and social responsibility, the work done in foreign countries on adult education, the educational requirements of community, continuing education, vocations, agriculture, training, liberal teaching, evening colleges, correspondence courses, adult education and related disciplines, etc.

Education Commission's Role

Adult education is one of those aspects of education on which the Kothari Commission has thought at length. The Education has expanded the scope of adult education, and for its success and spread, it has suggested the following six-point programme:

1. Eradication of illiteracy.
2. Continuing education.
3. Correspondence courses.
4. Libraries.
5. Role of universities in adult education.
6. The organisation and administration of adult education.

We will now consider the recommendations of the Commission in the light of its programme.

Educational Movement

The Education Commission has pointed out that as compared to the situation in 1951, there were 26 million more illiterate people in 1961. As compared to 1951, there were 20 million more illiterates, despite the fact that literacy had increased from 16.6 to 28.5 per cent. It also accepted the fact that the nation was paying a heavy price for this illiteracy, because illiterate persons are not really independent and complete citizens. Besides, illiteracy acts as an obstacle in economic, social and political development, control of population, creation of security, etc.

The Commission has suggested that any programme for the eradication of illiteracy should have the following bases:

1. As far as possible, adult education should be based on work, and it should aim at change of attitude, creation of interests and skill. The information given should bring about an increase in the learner's skill.
2. Adult education should stimulate the learner's interests in national problems and also make a powerful and effective contribution to the nation's social and political life.
3. It should make adults capable of reading, writing and calculation. If the learner's wish they can ensure their own improvement through informal knowledge.

It is evident from these three bases that the literacy drive will have three stages:

1. In the initial stage, the adult should be taught reading, writing and calculation, and also given information about citizenship and national problems. Besides, information about his vocation should also be imparted to the learner.
2. In the second stage, the knowledge obtained in the initial stage should be reinforced and the skills developed further. Adults should be enabled to understand and face their personal problems.
3. The third stage should help the adult to participate in continuing education.

A useful programme of adult education depends upon the acceptance of certain basic facts. We must see such programmes in the light of industrialisation and modern agricultural progress, as making illiterates useful instrument in the country's general as well as economic development, as turning them into an effective workforce. For this, the following facts should be considered:

Targets: The success of adult education depends upon successful planning. There should be a proper time schedule for programmes for the people, preparation of resources, training of workers and provision of other needs. The Commission has recommended that the programme should not be prepared for the whole nation, instead, state-wise planning should be adopted. The progress of adult education depends upon the educational facilities, public cooperation and existing organisation in each region. The Education Commission has expressed the hope that, on the basis of proper planning and implementation, 60 per cent illiteracy can be eradicated by 1971 and 80 per cent by 1976. Consequently, the Commission has recommended that every effort should be made to eradicate illiteracy. It has hoped that within 20 years, no part of the country should remain backward.

Programmes for Eradication of Illiteracy: The Commission has suggested the following programmes:

1. Providing for compulsory education of at least five years for the 6-11 years age-group.
2. Providing for part-time education of the children in the 11-14 age-group who had to give up their education for some reasons.
3. Making arrangements for the part-time general as 'well as vocational education for individuals in the 15-30 age group.

This group will include individuals who have received education partially.

Strategy of Approach: The Commission has rightly argued that every effort must be made to overcome obstacles to adult education. We must consider the population in our country and determine our strategy accordingly. According to the Census of 1961, there were 1 million illiterates in the 15 + age-group. Literacy in urban areas was 47 per cent, while in rural areas, it was 17 per cent. The distribution of literacy also depends upon the region. While in Delhi it was 52.7 per cent, in NEFA it was as low as 1.8 per cent. Besides, there was considerable differences between the literacy among men and women. Social groupings, caste and class and the nature of the community have always hindered adult education. Besides, the differences in the nature of a region lead to different levels of motivation towards adult learning. The Commission has felt the need for the acceptance and observation of certain general principles, and so it has suggested two approaches for success in this endeavour:

1. Selective approach.
2. Mass approach.

Programmes based on both approaches should run simultaneously.

According to the Commission, the selective approach is suitable for the group which can be identified easily and also kept under control with relative ease. Such a group can be effectively motivated to obtain education. Such a group can be made useful and the literacy programme can achieve success. Even vocational training can be given to such groups. Under this approach, the following programmes can be formulated:

1. Almost 4 per cent illiterate people are employed in industrial and professional organisation. Hence, the Commission has recommended that large firms, trading and industrial organisations and institutions may be legally made responsible for the adult education of their employees. They should be made literate within three years. The owners of such organisations must bear the entire responsibility of making their employees literate. The government should bear all expenses related to education, such as supply of teachers, teaching materials, books and other tools. The Commission is convinced that the owners themselves will, in the long run, benefit from the education of their employees.
2. The Commission has suggested that large organisations managed by the administration must give primacy to the important programme of adult literacy.
3. The humanitarian attitude is necessarily linked with all programmes for social and economic progress. They include a large number of people who have never even seen a school. Hence, it appears logical to argue that provisions must be made for the education and training of all workers in the schemes for industrial, agricultural, and trading development or improvement in health and education.
4. The Indian government has implemented numerous schemes for social welfare and economic development. These include such schemes as the creation of the Khadi and Village Industries Commission, community development schemes, child welfare schemes, etc. The Commission's view is that adult education should be made an integral part of all such programmes.

In order to put an end to illiteracy, it is necessary to mobilise all literate people in society. This approach is unorthodox, and it has not yet been adopted in practice in India. Russia has achieved spectacular successes in adult education through this approach. In India, Maharashtra has achieved notable success through its Village Education Muhim. In this approach, the workers seek the cooperation of the community and make schemes for mass education. The only difficulty in this approach is the laxity in follow-up, but this can be taken care of. Such a programme possesses the following characteristics:

1. This approach depends more upon political and social institutions rather than administrative educational institutions.
2. Adult education, by its very nature, is voluntary, and the drive behind it is the individual's own motivation. Hence, national unity and security, health and social welfare should be implicit in adult education, a fact which must never be lost sight of by educational planners, administrators or workers.
3. This approach depends upon the voluntary services contributed by such educated individuals in society as doctors, lawyers, teachers, students, engineers and other individuals, though, of course, the brunt of the responsibility falls upon the teachers and students. It is for this reason that the Commission has recommended that adult education should be propagated, through compulsory national service at every level of education.
4. In connection with adult education, a new responsibility should fall upon the schools. This will also help to modify the attitude of the school. This approach comprehends not only the student in the school, but the entire community. It provides an important sphere for extension service.

A programme of adult education cannot be run successfully as long as a proper plan is not formulated. For the success of such a programme, the following points must be considered:-

1. Before the implementation of the programme, the government, and all social and political leaders should arouse the interest of the community and bring about mobilisation.
2. The individual adults taken into the scheme should be psychologically motivated in a suitable manner. They should become aware of the importance of literacy.

3. All means of mass communication, such as radio, television, films, slides, drama and literary organisations should be made use of in such programmes.
4. Whenever the programme is to be implemented, the necessary teaching aids such as books, charts, graphs, guidebooks, etc., must be prepared beforehand.
5. In planning for literacy local, needs and conditions should be kept in mind.
6. The literacy programme should also have provision for the continuing education of neo-literates. Follow-up programmes should also be organised.
7. The adult education programme cannot be left entirely and exclusively to the teachers. They must be offered the cooperation and assistance of the following: i) extension services, agricultural and industrial departments, and ii) the use of mass media instruments, such as the All India Radio.
8. Libraries should also extend their cooperation to this programme.
9. Training in leadership should also be part of this programme.
10. Students, teachers and others interested in the spread of adult literacy should be given suitable training.
11. A powerful administrative and supervisory mechanism must be evolved.
12. In this progamme, schemes should be devised for bringing people into contact with each other through study groups, committees, clubs, recreation groups, etc.
13. The public, too, should recognise and appreciate such programmes.
14. The percentage of illiteracy among women is greater than among men. In the urban areas, 34.5 per cent women are literate while in the rural areas this percentage is 8.9, a situation similar to that found existing in the whole world. Hence, as many lady teachers as possible should be appointed to speed up the education of women.
15. The success of any activity depends upon the follow-up action given to it. Hence, for adult literacy also, properly planned programmes of follow-up action should be devised.

Doorstep Education

Adult education occupies the most important place in our country's educational set-up. In the rapidly changing circumstances of today, life as a whole should be conceived of as a continuing process of learning. It is now being generally accepted that the modern method of education does not provide extensive education for one's entire life. Hence, to complement it, full-time, part-time or in-service educational programmes are devised for those who have left school and reached adulthood: Generally, the system of continuing education is devised for two classes:-

1. Those people who participate in part-time education along with other individuals. Individuals of this class increase their qualifications by receiving part-time education in schools and colleges.
2. Those persons who can study only at home, but who need some assistance in getting official recognition for their study.

Continuing education is meant for persons who want to complete the education left incomplete by their leaving school. Individuals working in institutions also need opportunities for improving their qualifications and abilities. There are also some individuals who wish to study literature, language or some special subject merely for the pleasure of it. Adult education seeks to bring education within the grasp of every kind of person by catering to his special needs.

Hence, the Education Commission has suggested the following steps for developing continuing education: 1) For part-time education, a parallel system must be developed within the educational structure. Just as degrees, diplomas and certificates are granted to full time students, similar recognition should be available for part-time curricula and courses also, 2) Education institutions should provide the necessary leadership in this process; they should understand the individual's problems, solve them and help them in obtaining knowledge as well as the necessary help, 3) Through special part-time courses and mixed programmes, efforts should be made to improve the mental horizon of workers and employees; their knowledge and skills should be improved and a sense of responsibility engendered in them, and 4) Specialised institutions should be established, for example, the Central Women's Welfare Council, Vidyapeeths, etc.

For millions of those people who have to stand upon their own feet very early in life and struggle for livelihood and survival, correspondence courses provide a means by which they can satisfy their ambition for education, which would otherwise remain unfulfilled. Such courses are very much in use in foreign countries, such as the USA, Sweden, Japan, the USSR, Australia, etc. In India, too, experiments have been conducted in this form of education by the Delhi, Meerut and Himachal Universities. Although this method does not provide any immediate and direct contact between the students and the teachers, the relations between the two can become quite intimate through exchange of correspondence. Besides, it is also to be noted that there has generally been a dearth of teachers who can inspire students through direct contact.

Correspondence courses should be designed to encourage an exchange of views, and they should not restrict themselves merely to the exchange of written information. Individuals who wish to study science through the correspondence courses should be allowed to make use of laboratories during vacations. Even in education as a subject, there should be arrangements for correspondence curricula through Programmed Learning, a method which has been productive of good results.

The correspondence system of education should also get the assistance of radio and television, which can be used for the expression of fundamental and emotional ideas. Courses of this kind should be designed not only to enable students to obtain degrees, but also to give information and knowledge to individuals engaged in agriculture, industry and other productive activities. These courses should be organised through the extension departments.

The following are the recommendations of the Education Commission for the development of correspondence courses:

1. Those individuals who can take part in courses or curricula should receive education through correspondence courses. For this, extensive correspondence courses should be organised.

2. Students studying through the correspondence system should be given opportunities to meet the teachers at regular intervals. As far as possible, they should also be allowed to avail of library facilities existing in schools and colleges.
3. Correspondence courses should be assisted by the radio and television network.
4. Correspondence courses should also be organised for those individuals who wish to develop their literacy or aesthetic values.
5. Correspondence courses should not limit themselves to helping students merely to obtain university degrees. Instead, they should also contain courses which may lead to increased production in agriculture and industry.
6. Through correspondence courses, school teachers should come into contact with fresh knowledge, new methods of teaching, etc.
7. The education ministry, with the assistance of other ministries, should set up a National Council of Home Studies. Its functions would be to evaluate the correspondence courses, the agencies involved in such education, and the identification of spheres in which correspondence courses can be introduced.
8. Those individuals who wish to appear at examinations conducted by Boards or universities, on the basis of self-study, should be allowed to do so.

Libraries have always played a leading role in the growth of knowledge. In their absence, the possibilities of developing the habit of reading are very few, especially in rural areas. The Commission has suggested that, for the spread of mass education, school libraries should be utilised as public libraries. It has suggested that:

1. The suggestions of the advisory committee on libraries should be accepted, and on its basis, a network of libraries should be spread over the entire country.
2. School libraries should be organised and structured as public libraries.

3. Libraries should be dynamic; they should attract adults towards themselves.

The importance of libraries is reflected in the following statement from UNESCO. It says that in the whole world, and especially in developing countries, there is great need for the expansion of free libraries as a part of the educational programme. Libraries satisfy the needs of the prevailing educational system and help the teachers in their work. In the absence of public libraries, many adults and young men gradually sink towards illiteracy on leaving the school.

Role of Universities

Commenting upon the importance of universities, the Education Commission has declared that the image of a university is that of an intellectual, 'closed' community which gives rise to knowledge, spreads and preserves it, but this is the traditional image. The walls which divide 'town and town' no longer exist. There has come about a revolution in the life of the university, with the revolutionary idea that the community can be linked to the university for the mutual benefit of both.

The Delhi University introduced the correspondence courses and soon found that the demand for them was growing rapidly. In the same manner, the Rajasthan University set up a department of adult education, which was widely applauded. The idea was widely appreciated in Meerut University also.

The function of the university is to bring about a growth and improvement in the social, economic, educational and cultural spheres, and through special agencies it can have a profound impact upon these spheres. Universities should organise effective and powerful programmes for various vocations. It is their duty to ensure that their teachers are constantly in touch with the latest editions to knowledge. In the context of adult education, the Education Commission has made the following recommendations regarding the role of universities:-

1. Indian universities should shoulder a heavier burden in educating the adults in the country. This can be done in the following ways:

a. Departments similar to the Rajasthan University's department of adult education should be set up in other universities.

b. Universities must contribute to social and economic growth.

c. They must nourish community life and cultural values.

2. A Board of Adult Education should be constituted by taking representatives from the various departments involved in making the adult literacy programme a success.
3. Universities should be given financial aid for the proper implementation of the adult education programme.

Government and Education

The Commission has clearly felt that the basic weakness in the efforts being made by governmental, non-governmental and voluntary agencies towards the spread of adult education lies in the absence of proper planning and cooperation. Hence, it has suggested the establishment of a National Board of Adult Education to overcome this shortcoming. The council should have representatives from all related ministries. It should have the following functions:-

1. Advising the central and state governments on informal adult education, offering for consideration blueprints for teaching and training.
2. Wherever necessary, setting up institutions, helping in the preparation of literature, teaching materials and necessary teaching programmes.
3. Creating harmony between the efforts of various departments and governmental as well as non-governmental agencies.
4. Evaluating the achievements from time to time and making suggestions for future improvements.
5. Encouraging research, evaluation and survey activities.

Similar organisations should be established at the state level also. At the district level committees should be set up to function as integral parts of the district councils. They should be assisted by village and block committees. They should also be in intimate contact with extension programmes in the spheres of agriculture, cooperatives, health, community development, etc.,

On the question of administration, the Commission's view is that a separate department within the ministry of education should organise this work. The government must accept full responsibility, and every department of the government should be cognizant of the importance of adult education. In addition, every department should help in implementing this programme. Undoubtedly, adult education is the concern of the department of education, but it is also true that, in order to achieve success, it will be necessary to depend upon some particular mode of action.

As far as the question of voluntary agencies is concerned, they should be given financial and technical assistance. Adult education is a sphere in which such agencies can work freely. Besides, the nature of the work is such that its progress and mobility provides a necessary background for the success of all our planning for national development.

Fast Progress

During this period, nationalist governments were set up in the provinces. Dr. Sayad Mahmood, the education minister of Bihar, himself went to villages, equipped with a blackboard and chalk, to educate the villagers. Dr. Rajgopalachari himself wrote textbooks and ran a strong movement for literacy over a period of three months. Teachers of primary schools made valiant and commendable efforts to spread literacy in villages, and as a result, 12,000 people were made literate within these three months. In this way, for the first time the responsibility for adult education was grasped by the Indian mind, and the government took this onerous task into its own hands. New curricula were formulated. The medium of teaching was changed. Books, posters, film shows and other things were recognised as instruments of adult education.

At this time, literacy movements were launched in Assam, Bengal, Bihar, Uttar Pradesh, Orissa, and Jammu and Kashmir, keeping in mind local conditions and needs. New textbooks were prepared for the neo-literates in regional languages, and a separate department for adult education was formally created. After 1942, the urge towards education and literacy grew even stronger in the country, the outcome of which was that by 1947 as many as 9,50,002 adults were made literate, and 6,33,896 books were distributed among them. The Talukka Library and the Bombay City Adult Education Committee were set up in Bombay, An Education Expansion Department was established in Uttar Pradesh, while comparable organisations quickly sprang up in other parts of the country also.

The period 1942-47 was a very difficult one for adult education because it was buffeted on the side by the nationalist movement, and on the other by the instability and indecision of the government. Despite this, the movement continued because the people were keen to receive education.

After independence a central education advisory committee was constituted. The committee then set up a council to reflect upon the potential for adult education. This council took adult education out of the limitations of a knowledge merely of letters and raised it to the level of authentic knowledge, and, as a result, adult education was rechristened social education. A plan for social education was put before a committee of State ministers of education on February 19-20,1949, and the Government of India distributed Rs 60 lakhs to the States for adult education.

Social education centres were established in Bihar, where, in order to speed up the pace of social education, such programmes as touring drama troops, troops of folk dances and hymn singers, etc., were organised. In Madras, libraries were developed, and a training camp for teachers was organised under the auspices of social education. Similar work was also done in Bombay. In Madhya Pradesh, literature was prepared in Hindi and Marathi. In Uttar Pradesh, Etawah a project was initiated. It was later assimilated with the development programme and social education officers came to be appointed. The Government of India, too, has played an active role through such organisations as the Institute of Audiovisual Education, the All India Radio, the Institute of Social Education, which became involved in preparing literature and discovering new techniques of adult education as well as the training of teachers.

The Kothari Commission, in particular, is extremely conscious of the need for and the importance of adult education. In one of its statements on this subject, it says that though, to its observation, the history of adult education in India is not an old one, literacy had risen to 29.3 per cent, a fact which clearly implied that the movement should be suitably accelerated. In the sixth plan, the programme of adult education is being implemented on a national scale.

Targets and Aims

Adult education must be looked at from a new perspective. In planning for education in backward areas, it must be remembered that the adult has already gained experience of life but in receiving education he consciously regresses to his childhood stage. At the same time, he has many memories of family responsibility and of his contribution to the economy of the family. He has to discover and form social, political, constitutional and mutual relationships. Though he may be illiterate, these elements form his background. He does not possess extensive knowledge of society, but he has the requisite ability to learn all this. He can easily learn reading, writing and counting. Hence, it is essential that the adult should be properly understood before an attempt is made to teach him.

Social education is corrective or remedial education. Those who never got a chance to go to school, or could attend school only for a short period, find that their social education has remained incomplete. The deficiency can be made up through social education.

Today, many terms have become prevalent in the sphere of social education:- Adult education, social education, informal education, continuous education, etc. Each one of these terms is indicative of adult education, but they differ conceptually from each other. The general and obvious meaning of adult education is the teaching of adults. Bryson has said that education, in every situation and in every condition for the individual is what constitutes adult education. Organ and Bundy define it as the adult individual's conscious attempt to learn. Rheins, Fenster and Hadley believe that adult education implies the education related to the individual's professional, individual and social life. The concept of a social education is a more comprehensive one in as much as it comprehends reading, writing, counting, etc., and even more, becoming a skilful and useful citizen. The term social education has been coined because the purpose is to arouse civil awareness in the individual. Humayun Kabir believes that it is a means to develop civic consciousness and social solidarity. In it, attention is paid to literacy, the creation of an educated mind, and the growth of a powerful citizenship for the nation.

S.N. Mukerjee: Social education is roughly inclusive of all the informal education that is given to adults. In India, it has two aspects: i) adult literacy, i.e., the education of those adults who have not received any kind of education in schools, and ii) continuous education for literate adults.

Barker: His view is that adult education constitutes that education which an individual receives on a part-time basis while continuing in his occupation.

Bryson: According to him, adult education comprehends all those activities which are related to the normal life of the individual, which have some importance for education, in the doing of which the individual has to utilise only one part of his total intellect..

K.G. Sayyidain: He sees adult education as inclusive of education of politics, of citizenship and also of morality.

Humayun Kabir: His opinion of adult education is that it aims at the development of an educated mind in an illiterate individual.

Implied in the concept of social education is the fulfilment of the needs of adults-economic efficiency, increased knowledge and the growth of cultural possessions. The objectives of social education can be pinpointed as follows:-

Remedial: Here, the purpose is to increase the individual's original or innate talent.

Vocational: In this case, the purpose is to impart commercial and industrial knowledge in urban areas and knowledge pertaining to agriculture and domestic activities in rural areas.

Health: Social education aims at acquainting adults with the basic principles of good health and the various means of maintaining hygienic conditions.

Social Skill: The objective of social education is to train the individual in maintaining harmonious relations with friends and companions, leading a satisfactory family life, and becoming conscious of one's rights and duties in society.

Recreational: In order to raise the mental level of adults, adult education seeks to develop the habit of healthy recreation. The varied means of entertainment available today have not had a desirable impact upon the minds of the people, but healthy recreation helps the individual develop his life along the right path.

Self Development: Social education aims at enabling the individual to bring about his own harmonious development by developing a thirst for knowledge, forming an individual philosophy of life and evolving a creative and aesthetic approach to life. Thus, it becomes apparent that the foregoing objectives of social education fulfil four important goals of our- national life:-

Growth of Civic Sense: Through social education, those who have been left behind in the race of progress are enabled to overtake others and to march in step with them, thus, being enabled to perform their tasks with the cooperation of others.

Utilisation of National Resources: Having been influenced by adult education, adults are enabled to fulfil national objectives in their own respective spheres. Lenin said that socialism can never come through an illiterate populace because the illiterate is isolated from politics. Lenin's opinion may, or may not, apply truthfully to socialism, but it is certainly an axiom in the field of good citizenship.

Creation of Cooperative Associations: Through social education, the adult individual achieves economic as well as social progress, and thus becomes enthusiastic about the creation of cooperative associations.

Growth of a Social Viewpoint: Society as a whole cannot evolve a particular social viewpoint if its members are uneducated. For instance, the realisation that "we are Indians" can be developed and rooted deeply only when the people are educated.

On the whole, it can be said that our country needs adult education for a social and political revolution. It can, on the one hand, develop emotional and national unity among the people, and on the other, develop self-confidence, fearlessness and skill in the individual. In fact, the success of democracy depends, to a very large extent, upon social education.

Different Streams

Social education is the responsibility of society. In a democracy, every educated individual contributes to the shaping of society in his own individual way. And, the greater the extent to which society shoulders this responsibility, the better is it for society itself. Thus: a) the government, b) private institutions, and c) universities and training institutions contribute to the spread of social education. At the governmental level, many ministries perform this task through a variety of programmes. Social education is imparted to the masses through: i) correspondence courses, ii) vocational adult education, iii) training of adult educators, iv) agricultural and other professions, v) public libraries, and vi) museums, etc.

Values Applied

Social education is an educative process that continues through life, and hence its field is very vast. It is inclusive of spread of literacy, knowledge of rules pertaining to the safeguarding of health, vocational training for improving economic condition, feeling of citizenship, awareness of rights and duties, and healthy entertainment. In determining its scope, the following points are kept in view:-

1. Social education comprehends those subjects which are not taught in schools or those which are neglected.
2. It brings about a change in human attitudes.
3. It either reduces or eradicates the differences in social and individual progress, something which should be done through: a) knowledge, b) deeds, and c) determination.

Subject's Importance

Every individual today is a useful part of society. The purpose of social education is that the adult population should be educated in such a way that it may be able to improve its own life, develop the desire to create a better and a modern society instead of adhering to the traditional society, and also have faith in, the future of its nation. The responsibility for the field work in social education, including adult education, rests with the state governments and the administration of union territories.

In view of this, the importance of social education can be specified in the following terms:

1. Social education provides the foundations for the success of democracy.
2. It is the lamp of hope for the uneducated.
3. It brings about the political, intellectual and social development of adults.
4. It inspires the aesthetic development of adults.
5. It strengthens the cultural aspect of community life.
6. It prefers new media of expression and forms of recreation.
7. When all other facilities are available in society, the existence of illiteracy is a blot.

Planning Regularly

In the first five year plan, the education ministry provided financial aid to various projects to ensure the progress of the social education programme. Ideal community centres were set up and some special primary schools were given the form of School cum-Community Centres. Under this scheme, 116 Modern Community Centres and 454 School-cum-Community Centres were created, in addition to which many libraries, rural libraries, service centres and public colleges came to be established.

During the second plan, the National Fundamental Education Centre was set up with the assistance of UNESCO. Its objectives were the following:

1. Training workers of noteworthy importance for social education, for example, district social education organisation.
2. Conducting research.
3. Creating audio-visual aids and teaching materials.
4. Performing the functions of an information centre.

At the same time, the Social Education Institute was established at Indore. New literature was produced for the neo-literates and literary camps were organised. The Delhi University and the National Book Trust undertook the task of training librarians to man public libraries.

By the third plan, literacy had risen from 16.6 per cent to 24.0 per cent. In different states, different kinds of programmes of social education were implemented. In addition to the public or Janata colleges, six Vidyapeeth institutions were set up in Mysore State. A social education department was set up in Rajasthan University. 205 district libraries were created in 13 States, and States also gave financial aid for distributing books in rural areas.

The three preceding plans had not given requisite importance to social education. In the fourth plan, a sum of Rs 3.6 crores was allocated by the Central government for this purpose. The sum was invested on the following activities:-

1. Organising and strengthening the social education administration.
2. Establishing labour Vidyapeeth institutions.
3. Creating suitable literature for the neo-literates.
4. Developing and expanding the library service in Delhi.
5. Developing a central institute of library science.
6. Training librarians.
7. Implementing other services related to libraries.
8. Giving aid to voluntary educational institutions.
9. Initiating practical schemes, setting up Janata colleges, and organising continuous education and adult schools.
10. Developing social education institutions for the workers.

New literature was got prepared for the neo-literates with the assistance of UNESCO. 419 books were given awards, and an adult encyclopaedia was published under the title *'Gyansarovar'*. A book entitled *'Hindi Vishwabharati'* was published in ten volumes. Workshops were organised as a part of literary camps. Among the non-governmental organisations which do praiseworthy work in the sphere to social education are Indian Adult Education

Association, Social Adult Education Committee, and the State Adult Education Council.

The Central Education Advisory Council, at its 31st meeting, laid down the following standard or criterion for adult education:-

1. Recognition of about 350 words.
2. Ability to write numerals upto 100.
3. Ability to read simple charts, posters, etc.
4. Knowledge of objects of daily utility.
5. Ability to write simple sentences and names.

Again, relatively less importance was attached to social education in the fifth plan. The movement that was initiated in 1937 for the total eradication of illiteracy among the masses remained a failure then, and because of this, adult education came to be modernised in the form of social education. This programme was implemented throughout the country through the medium of community development programmes, but, for a variety of reasons, even this programme remained ineffective. As a consequence, the social education programme continued to function primarily through local institutions, with the assistance received from the State and Central governments.

However, many important ideas and schemes came to light. The Vidyapeeth of Mysore, the Gram Shiksha Muhim of Maharashtra, the Bombay City Social Education Committee, the Literacy House of Lucknow, and the Rural Library Movement of Lucknow, etc., are the main forums of social education. The Farmers' Education and Creative Literacy Movement initiated in 1961 has proved of great significance. By the end of the fourth plan period, it was expected that ten lakh farmers in 100 districts should have benefited from this movement. In addition, departments of adult education were set up in some universities for giving impetus to continuing education. Through social education, efforts were made to provide information about the various aspects of life through programmes concerned with agricultural extension, handicraft training, family planning, child and family welfare, nutrition programmes, etc. However, the total impact of these programmes has not succeeded in bringing us near to the national goal of eradicating illiteracy.

With the expansion of primary education, the percentage of literacy, which was 24 in 1961, rose to 29.6, but the equally bitter truth was the fact that the number of illiterates rose from 333.6 million in 1961 to 386.7 million in 1971, which was obviously due to the rise in population. However, if the age groups 15 to 25 years or 25 to 45 years are taken into account while considering the above statistics, the situation appears to be a little better.

For the fifth plan, the allocation of funds for adult education is illustrated in the following table:

Adult Education Plan

Plan	*Expenditure Rs. in Crores*
1. State Adult Education Council	1.00
2. Literature for adults	2.00
3. Functional Literacy	4.00
4. Aid to Voluntary agencies	1.00
5. Research and Experiments	2.00
6. Capable Schemes	10.00
7. Continuing Education	30.00
Total	50.00

Continuing education is that form of adult education, or social education in which educational opportunities are provided to those who have given up education prematurely. In the sphere of higher education, continuing education is required most in the sphere of technical education, a fact indicated by the surveys and conclusions of the Scientific Manpower Committee (1949), Engineering Personnel Committee (1956), the Working Group on Technical Education and Vocational Training (1960), and the Technical Manpower Assessment Committee (1966), etc. Technical education points to the path for the future plans. Hence, it becomes necessary to provide opportunities for education through correspondence courses, part-time courses or self study courses so that abilities may be developed and people may remain in touch with the latest addition to human knowledge. For this purpose, it is desirable to formulate a general policy.

Considerable thought was given to bringing adult education into the list of the nation's priorities. In the first place, social justice cannot be made available, and democracy cannot succeed in the absence of training in productive techniques, knowledge of fundamental economics and creation of the right attitudes. And, the success of democracy and social justice were the foundation stones of the fifth five year plan. And it was generally accepted that even the success of such a vital programme as the Green Revolution depended upon the success of adult education.

The various experiments conducted so far in social education have brought to light many important activities and ideas. All such programmes must have the inbuilt quality of attracting adults towards themselves. There is necessarily an intimate relationship between the adult's routine activity and social education. It should be closely related to introductory education, expansion of agriculture, cooperation, etc. Social education must be linked with the national service scheme, and even the predetermined curriculum will have to be linked to national as well as local requirements. It is also inseparably related to research and the creation of suitable literature. We must lay down a variable network of libraries in rural areas, and our aim should be to make illiteracy a thing of the past within the next few years.

1. Linking creative literacy with such nationwide and productive programmes as the extension of agriculture, animal husbandry, reclamation of land, irrigation and drinking water schemes.
2. Linking creative literacy with such important national programmes as family planning, child and family welfare programmes, etc.
3. Initiating programmes for health and nutrition, literacy, recreation, social organisations, etc., wherever the density of population is high.
4. Laying stress on the programmes that proved successful in the past; organising adult schools and polytechnics, etc.

Of the allocated sum of Rs 28,750 crores for social education, 20 per cent was spent upon these programmes. Rs 580 crores were to be spent upon the training of employees. According to the plan, the States and the Centre could make provisions for Rs 100 crores, while Rs 20 crores were set aside for setting up departments of continuing education in various universities. At the same time it was also true, that, of the total educational budget of Rs 2,200 crores not even Rs 120 crores could be set aside for social education, and hence a provision of only Rs 50 crores was made for this purpose.

Adult education has been given national importance in the blueprint of the sixth five year plan. In India, about 20 crore people are illiterate. In addition, about 70 per cent of the children opt out of schools by the time they reach the fifth class, and so they also fall within the categories of illiterates or semi-literates. For this reason, the sixth plan lays special emphasis upon the eradication of illiteracy, and creation of universal primary education and industry-centred education. There is a scheme for introducing a nationwide educational programme for the illiterates in the 15 to 35 years age group. For this, it is intended to seek the cooperation of universities, labour unions, public service planners and voluntary organisations. It is hoped that the scheme will benefit 10 crore individuals.

Under this plan, importance has been given to the speeding up of earlier programmes, which include: i) creative literacy of farmers in rural areas, ii) setting up of labour schools and multipurpose adult education centres in urban areas, iii) search for adult education centres in universities, setting up of Nehru Youth Centres, and iv) national service scheme.

The plan aims at activating all these schemes, and it is hoped that literacy will be brought to 15 lakh people in the first year, 45 lakhs in the second year 90 lakhs in the third year, 180 lakhs in the fourth year, and 320 lakhs in the fifth and final year of the plan. This objective can obviously be achieved only through a powerful, effective and committed administrative mechanism. Some of the special features of this plan are as follows:

Attention to Weaker Sections: The plan provides for the education of the weaker sections in rural as well as urban areas, with special attention to providing literacy to the people in the 15 to 35 age-group. Special attention will also be given to women, Scheduled Tribes, landless labourers, farm labourers and other weaker sections of society.

Society and National Development: The plan envisages that special efforts should be made to ensure that the poor and illiterate individuals develop the necessary awareness about themselves, become conscious of the social reality surrounding them, organise themselves to overcome the difficulties of daily life and become partners in the process of social and national development. Apart from literacy, attention will also be paid to the fact that mixed programmes catering to the needs of the poor are initiated, comprising of general education, health education, family planning, development of vocational skills, science in relation to daily life, knowledge of technology, physical education and cultural activities.

Inspiration to Adults: The success of the adult literacy programme depends upon the motivation of the adults, and for this purpose particular attention is to be devoted to the selection and training of workers, production of good quality teaching materials, introducing developed methods of learning through the medium of doing and living, surveys and facilities for guidance.

Cooperation of Voluntary Organisations: In order to make the plan a success, the cooperation of voluntary organisations, youth and others interested in social service activities as well as organisations engaged in commercial activities will be sought. Labour unions, retired teachers and those engaged in formal education will also be made use of as far as possible, and adult education will, again as far as possible, be linked with other developmental activities envisaged in the plan.

Stress on Decentralisation: In order to make adult education a success, emphasis has been placed on decentralisation in the sixth plan. A National Board of Central Education has been established at the centre, and it is planned that similar agencies will be created at the state, district, block and local community levels.

Programme: In the sixth plan, the programme for adult education is spread over the entire plan period. The first year is to be devoted to preparation. It is hoped that this programme can be given suitable momentum by making schemes at the block and district level after consulting all related agencies, motivating voluntary youth workers, retired employees, other agencies, the adults themselves, establishing resource centres in every region, preparing suitable curricula, preparation of teaching materials, devising methods and materials for training, training of instructional personnel at the district and project level, etc. All these activities can be regarded as preparation for the programme.

Cooperation Schemes: In this programme attention has also been paid to the state agencies, centres for national development, industry, service planners, entrepreneurs, labour organisations, and the specific qualities of particular areas and groups. The programme has been kept free from the formal modalities of the department of education. In order to achieve the desired progress, the administration, cooperative units and the inspection machinery have also been mobilised.

Finance for Adult Education: In view of the national importance of the problem of adult education, a sum of Rs 200 crores has been allocated to it. This amounts to 10 per cent of the education budget, whereas in the fifth plan, the allocation for adult education amounted to only Rs 18 crores, or 1 per cent of the education budget. This budgetary provision is only one source, and it is planned that other resources will also be exploited for this purpose. For example, the various service planners and project officers will provide finance for their own adult education programmes. In the tribal regions adult education will also be financed from the budget for the agricultural and rural development, there will be provisions for providing functional literacy to the agriculturist. Thus, the provisions for adult education within the total educational budget is only one of the provisions. Once the scheme is implemented and experience gained, more funds will be made available, as need arises.

Community Factors

The leadership is not merely a personality characteristic, neither can it be understood exclusively in terms of the leadership tasks just alluded to. Drucker says of an institution, that it is "like a tune; it is not constituted by individual sounds but by the relations between them." The same can be said of leadership; it is a function of the social relations expected of certain social positions. These expectations both support leadership actions and restrain them.

The content of the leader's orders is normally as important as he who gives them. For no matter who the leader is, his orders cannot be arbitrary, that is, outside the realm of normal expectation. For example, it has been found that small group leaders cannot maintain their leadership positions unless their orders are in agreement with the traditions of the group.

The view of leadership as a social relation directs attention away from the personal qualities of the leader and towards the fundamental nature of leadership, the demands made by one party on another. Leadership occurs when subordinates comply with an order. A leader's acts cannot be understood out of context of his relationship with his subordinates. The nature of that relationship is reviewed below.

The leader has a unique role with respect to his subordinates. It is a role which is indispensable to their jobs and to the organisation. It is characteristically a role on which others depend. This characteristic is a matter of degree and is present in many roles. Thus, a member of a group is a leader to the extent that others in the group must depend on him to fulfil his tasks. Leadership, in other words, is not an all-or-none category, but is a variable. People are not simply either leaders or followers. There are as many leaders of an organisation as there are shared dependency roles; the same person may lead and be led as he changes roles.

The criterion of a leader, then, is the fact that all other members must depend on the leader more than he depends on any one of them. Thus, while every member of the assembly line performs a minor task, the foreman is less dependent on a particular assembly worker than each is on him – for parts, for wages, vacations, and so on. As specialisation increases, there is opportunity for more individuals to become indispensable at various times during their careers; but conversely, it becomes increasingly difficult for a single individuals to be completely indispensable and dominate the leadership structure.

This element of shared dependency differentiates "small group" leaders from large organisation leaders. In the former case, group functions are limited and so there are only one or two leaders because all members can be served by a single person. But in complex organisations with diverse functions, it is difficult for any one or two persons to serve the entire membership. They have neither the skill nor the social capacity to do so. This difference makes it hazardous to generalise in both small and large group leadership studies.

Social Elements

Bennis has characterised leadership in operational terms as comprising the following fundamental elements:

1. an agent;
2. a process of inducement;
3. subordinates;
4. the induced behaviour; and
5. a particular objective or goal.

He explains that the process of inducement may be defined as power- that is, the ability to control rewards and punishments and thereby control the means for the satisfaction of subordinates' needs. The behaviour that is induce is defined by Bennis as influence. Putting these elements together, leadership may then be defined as "......the process by which an agent induces a subordinate to behave in a desired manner."

Interaction and Informal Leadership: Like the ancient alchemists who sought a man-made substitute for gold, administrators understandably seek a set of rules, or "principles of administration" which apply unconditionally to most situations. There are at present no such rules. However, also like alchemists, the belief in them causes some persons to glibly accept the prescriptions of "more experienced" persons on faith. One writer cautions that it would be dangerous if there were such rules because, although they might work for a time, they would become inappropriate under new conditions. This in fact is the epitome of "trained incapacity," this is, the substitution of rules for aims. Yet this is precisely what the administrator who wishes to learn only from experience seems to be seeking.

Rather than look for a set of administrative rules, the administrator is better advised to look for underlying principles of human interaction in large scale organisations which will provide a method for understanding situations in which the aims must be achieved. Although the search for rules of administration assumes that situations are consistent, the quest for principles of interaction starts with the variability of situations.

Homans chooses to investigate small processes in terms of four concepts: activity, interaction, norms, and sentiment. Activity refers to the tasks of the work group. Interaction occurs whenever one person's activity is stimulated by the activity of another; it includes verbal and non-verbal communication. A norm is a standard held by work group members that prescribes what ought to be done; orders are norms which anticipate changes in established ways of doing things. Sentiment refers to the subjective qualities of members, their likes and dislikes, drives, emotions, motives, and attitudes.

The underlying assumption of the conceptual system is that a change in one element modifies the others. Thus, for example, by increasing the group's activities, interaction will increase among the members, and in increasing interaction they are more likely to get to know one another on a personal basis, either as friends or personal enemies." "The more frequently persons interact with one another, when no one of them originates interaction with much greater frequency than the other, the greater is their liking for one another and their feeling of case in one another's presence." On the other hand, ".....the more frequently one of the two originates interaction for the other, the stronger will be the latter's sentiment of respect (or personal hostility) to ward him....." That is to say, hostility or respect, rather than friendship, is a more likely outcome when superiors interact with subordinates.

The concept originating interaction provide a clue to the detection of informal status relationships. Action is originated for another when he obeys requests, volunteers to satisfy the wishes of other, seeks information, and so on. Often persons who are not in official leadership positions originate interaction for others. The assignment of persons to high-ranking offices does not guarantee that they will necessarily exercise leadership initiative, for there are many factors in the official structure that may prohibit the exercise of initiative. Key subordinates may be in a better position to exercise unofficial leadership because of their close affiliation with the public, with pressure groups, or with influential community leaders. Leadership depends partially an access to crucial information about what is going on the group and within the community; this may involve access to files. For example, the school psychologist is in a position of leadership because he controls secret information about students.

Since intelligent leadership requires that outside influence be considered and used in making decisions, the leadership position will be one that provides a wide variety of outside contacts. This is, the higher the formal or informal social rank, the more interaction there is between that position and the outside groups. For similar reasons, informal leaders will have contact with more of the organisation's personnel than those who are not leaders, and consequently they are in a position to develop intimate knowledge of the group norms. That is to say, the leader's knowledge of group norms is no accident. It is gained through a complex network of communication with all members of the group. In other words, "the higher the person's social rank, the wider will be his range of initiation" within the group as well as outside it.

This suggests that informed leaders arise from those group positions which afford a range of internal and external contacts and that provide ready-made channels of communication. Because communication generally flows towards and away from the leader in greater volume than to other members, it seems reasonable that informal leadership develops at communication centres. Thus, the higher an individual's informal rank, the larger the number of persons for whom he will initiate interaction, in official rank. This high rate of interaction of informal peer group leaders with other group members facilitates the development of mutual sentiments; informal leaders, therefore, can be expected to rely more on consideration and friendship and less on deference, respect, and initiation than official leaders. The view that leadership is a social relations process calls for special consideration of sociological explanations.

The influence of the situation on leadership is unmistakable. Roosevelt and Hitler were products of their time who would not have risen to fame and infamy apart from the situation. The nature of the situation greatly influences the type of leader that will be selected. The shy boy may become an aggressive leader when an issue close to his heart arises or when his special skills are demanded; as one writer points out, the lowly army private may assume command under severe battle condition, when his lieutenant's skills are no longer sufficient, and the most flagrant criminal who has knowledge to survive under times of stress may be entrusted with leadership during a community disaster. It must be concluded that leadership is a combination of a situation, demand for special skills, and a sense of common destiny, as well as personal ambition.

Social Effects

The skills of leadership, like other skills, are developed through training and experience. However, because of the variety of situational contexts, it is by no means apparent what type of training and experience are most beneficial. However, there is reason to believe that, in face of the overwhelming variety of situations, the most effective training is at the level of principles of interaction and social analysis.

The highly specialised courses in administration, which grew like Topsy, do not seem to reflect these theoretical principles. Separate administrative courses usually offered for elementary, secondary, and other special types of school organisation did not emerge as a result of evidence that there are different leadership principles for each level of school organisation; they emerged instead from "experience" oriented assumptions that what is important to know are the system. But contrary to this, research evidence indicates there is more similarity between large organisations of different purposes than there is between small organisations of different purposes. Such characteristics as system size, school size, situational influences, complexity, and degree of bureaucratic administration perhaps provide a more meaningful basis of differentiating course work in administrative training programmes than grade levels when it comes to examining leadership problems in education.

The situational context also challenges the fetish of experience. Not only may experience in one situation be unrepresentative of another, but it may close alternatives and become so technique – oriented that the broader principles of administration are never fully realised.

From the situational viewpoint, the training programme is another "situation," and from that perspective it also influences leadership patterns. When for example, the training programme is closely geared, either officially or unofficially, to a particular school system, or when the training amounts to on-the-job training, the practical benefits are often more than offset by the lack of perspective and critical imagination which could be derived from systematic consideration of theoretical principles.

The recent trend towards theory development, utilisation of social science models, and the application of these to ease situations appears to be a move in the right direction. Training programmes which include intensive study in discovering the world of reality through social science techniques, and the application of their knowledge to analysis of the administrative world through case studies, field studies, and simulated situational materials are recent and promising additions to preparation programmes for administrators in many universities.

The Standings

Although the situational approach calls attention to the influence which the immediate setting has on leader behaviour, it still underestimates the extent to which social factors condition the leadership role. For beyond the official organisational setting is a cultural one, the outside society, which seriously influences the leader's internal functions.

Seeman broached the question of how the leader's social status influences his leadership roles in a sample of seventy-five public school superintendents in Ohio. It was found that their leadership style was affected by their conception of the place they hold in the community and in the general culture. Measure of status were designed to assess:

1. the leader's community status,
2. discrepancies between economic and social status; and
3. distortions between the leader's perception of his social status and his followers' perception of it.

The measures were compared to the subjects' ideologies of leadership and to subordinates' evaluations of their leadership effectiveness. Regarding status differences, when the leader gave himself high self-ratings on community status, he also claimed high responsibility and authority and tendency to delegate authority to his subordinates. Also self-rated, high-status leaders who perceived a significant difference between themselves and their followers were judged by subordinates to be receptive to change, communicated frequently with subordinates, and did not express "separateness" (social distance) attitudes towards subordinates. Moreover, those high-status leaders who perceived a relatively great status difference between themselves and their subordinates received favourable evaluations from subordinates.

Because this last finding seems to be contrary to the egalitarian concept of leadership, it bears further consideration. Given a democratic bias among subordinates and a demand for personal consideration, one might expect that subordinates would prefer school executives who are similar to themselves in community status. But the reason they do not is suggested in the same data, there is a positive correlation between "separatism" of superintendents as described by teachers, and percentage increase in teachers' salaries. That is, higher-status leaders were in a better position to enhance their subordinates' positions by virtue of their social distance, that is, the fact that they associated with community influential rather than with teachers.

Leaders who exaggerated their status in relation to teachers' placement of them, and who "underestimated" teachers' status (relative to teachers' self-placement), exhibited high separatism attitudes towards teachers and were unresponsive to change; these leaders received low ratings from their subordinates. In general, the findings suggest that a basic requisite of good leadership is the ability to accurately judge the attitudes of the total membership on relevant issues (though not necessarily on non-relevant issues).

The importance of social status appears to lie in the relative security it provides. When the leader is clearly of higher status than his subordinates, he is not threatened by them, and his response is both one of high communication and delegation of authority. Also, his community status places him in a more opportune position to support his subordinates. However, when his community status is not clear, either because of disparities between economic and prestige ratings, or because his own status aspirations create distortion, the leader is threatened by his subordinates. Under these conditions, his leadership style displays personal status enhancing and conserving tactics, which probably reduces his overall leadership effectiveness.

Besides the limited sample and its obviously exploratory nature, one major shortcoming of the study, recognised by Seeman, is that no attempt was made to differentiate the school leaders' status in the local community from his status in the general society. The findings may apply only to "locally-oriented" executives, whose reputation and status in the local community is important to them. To the extent that school executives are mobile, often marginal community members, operating in a nationwide or statewide framework, their overriding status identification may typically be to professional and national sources outside the community rather than to the local community. This reservation does not detract from the study's significance. However it is evident that training in "human relations" and "social skills" do not encompass the leadership task. Leadership training stops no shorter than consideration of the cultural context.

The Classification

The fact that the situation is important does not make personal background irrelevant, however, Personal career patterns (as opposed to personality traits), for example, constitute one aspect of the "situation". Most administrators of the largest and most influential school systems achieve their positions after a long succession of upward moves. The exact degree of mobility among chief school administrators is variable. Carlisle reports an eight per cent average turnover of local administrators. Turnover is inversely related to size of units; administrators tended to move because of desire for higher salary. In New Jersey, districts look for a new superintendent at least once every six years, while turnover in twelve mid-western states was found to be nearly twenty per cent annually; the highest turnover there was reported in the smallest districts, and the greatest stability was reported in medium-sized districts of 40 to 200 teacher?

Local and Cosmopolitan Leadership Patterns: Carlson notes several principal ways in which the patterns of leadership differ among local (place-bound) and cosmopolitan (career-bound) public school superintendents. Place-bound superintendents are "insiders" promoted from within the system; they constituted thirty-five per cent of a nationwide sample of 59 superintendents, while sixty-five per cent were "outsiders." Three types of outsiders were identified: hoppers, specialists and statesmen. Hoppers move frequently without benefiting their status and without leaving a lasting impression on the schools they serve. Specialists are hired to do a special job-financing or building and do leave a lasting impression, but go away when their job is finished.

The statesman leaves only when he feels he can do no more for the system. Largely because of what the statesman has been able to do for the system, the school board is usually satisfied with the system as he leaves it, and hires an insider as his successor. Insiders are often represented in larger systems. Outsiders are looked to for creativity, while insiders are hired to maintain stability. No insider reported that his board was dissatisfied with his predecessor. It would be difficult for the insider to effect change even if he wanted to. He would not be as likely to have the school board's support and he risks being identified, on the basis of his past and by virtue of his promotion, by the teachers as the "school board's man." If there is any creativity in such a system, it must come from the teaching profession rather than from the administration. Carlson concludes: "The insider adapts or modifies himself to fit the office; his performance adds nothing new to the role. It is not creative.

The place-bound superintendent seems to derive satisfaction from the office; he does not bring status to it. Coming from the outside, cosmopolitans are in a better position to bargain that are insiders, and accordingly they receive between 51,000 and 85,000 more a year than beginning insiders, and the insider never catches up. This demonstrates the crucial relationship that connections with the outside market have on the salary level. Persons not willing to enter the market or leave their present employment cannot command the present market salaries.

Both types of superintendents studied by Carlson engaged in rule-making upon taking office. Rule-making creates the impression that the successor is engaged in important activities, that he is going to "do" something, forcefully bringing to everyone's attention the fact that he is on the scene, in the process of making rules, the successor also learns who will support him on larger issues and who will resist. Insiders and outsiders, however, were concerned with different types of rules. Insiders made rules which essentially sustained the *status quo*, while eighty-five per cent of the outsiders' rules altered internal commitments or external ties of the system. Insiders' rules typically pertained to the technical-managerial facets of the school; for example, "All individuals are responsible for making classroom observations and follow-up conferences." But outsider's rules affected the institutional level of the organisation and changed its character; for example, the outsider might establish a kindergarten or employ social workers to serve the school.

When the successor was an outsider, the informal organisation also showed signs of change and realignment of conflict relations. One noticeable effect of the outsider was a temporary solidification of informal relationships within each level of the hierarchy; that is, interaction increased among elementary principals and among secondary principals, but decreased between them. Such solidification was less likely where the successor was an insider, and when it occurred it took place across hierarchical levels. This means that, unlike the insider, the outsider inherits strong. The insider knows who his friends are and who his enemies are, and he has had years in which to build up a personal following. Consequently, outsiders more frequently increased their staffs than did insiders. Even new outside superintendents added more positions than old outside superintendents, while the reverse was true for new and old insider superintendents, which indicates that additions by the outsiders are viewed as strategic replacements necessary in the early years. Thus, school systems changed more quickly when outsiders were brought in.

Insiders "lasted" in office longer than outsiders. The mean time in office for insiders was ten years, while outsiders lasted a mean average of eight years. There was some evidence that insiders were able to persist longer because of more thorough political adaptation to outside interests. One insider who served twenty-seven years in one system never permitted himself to take a position in conflict with his school board, but he was willing to sacrifice principle for expediency; he also spent his time in community projects making personal contacts.

Although there are periods when insiders are needed. Carlson concludes that "two insiders in a row may be one too many." Two consecutive insiders would mean that the system has endured an average of twenty years of leadership without any major adaptations to the environment, less than ten per cent of the instances of succession were concurrent insiders. However, about half of the cases were consecutive outsiders, fifty-three per cent of 209 instances were outside to inside, and thirty-nine were inside to outside successions. Carlson conjectures that by comparison with business organisations, which exist in a "wild" environment where adaptation is necessary, public schools are "domesticated" in the sense that they are assured of clients and income. They can exist longer without adapting.

Thus, it is apparent that turnover is a crucial aspect of the school's character, and that the type of person a board hires – outsider or insider – may be a more significant determinant of his leadership than personality traits.

This process of promotion, mobility, and aspiration to positions of greater authority affects the leadership style. In a study by Seeman of fifty school administrators, results of a leader behaviour description questionnaire developed by Halpin disclosed that while measures of career mobility showed little relationship to administrative behaviour, it was significantly related to a mobility aspiration scale designed to test the administrator's ambition to succeed. By combining mobility history with mobility aspiration, a four-way typology of administrators was developed as mobile non-strivers, mobile status seekers, stable non-strivers, and unsuccessful status-seekers. Both the mobile strivceptive, the mobile non-strivers displayed relatively uniform controls over the organisation and low responsiveness to needs of the group. The finding demonstrate that the career pattern and aspirations influence the style of leadership.

Aims and Factors

Another facet of the mobility pattern which conceivably influences leadership style concerns what leaders strive for. Some ambitious persons seem to be seeking to "become something," and are awed by the style of life or the prestige of the leadership position itself. On the other hand, some ambitious persons seem to aspire to leadership in order to *"do something"* – they want to effect a programme, or change one currently in effect. Conceivably, these two types of leaders display different leadership styles upon taking office.

For example, the "strong" leaders, who resist pressures that deflect from their goals, are perhaps more likely to be found among the latter type. "Weak" leaders, who succumb to power of outside interests or who allow subordinates to ultimately assume command, are conceivably of the former type; for having no programme, they seek out clues from others about how the role is to be played. In this case, it is the interplay between ambition and social pressures that constitutes the "situation."

Often, the personal goals that lead individuals into an occupation are not the ones that hold them there. One investigator classified nurses on the basis of shifts in their goals during their career: the "dedicated" had initially strong allegiance to nursing, which was retained throughout the career; the "convert" entered nursing with a low estimation of it, but radically altered her opinion of her career; the "disillusioned" entered the occupation with a high estimation of it, which later deteriorated The "uncommitted" had neither entered the occupation with a high estimation of it nor developed one during the career. In short, it is conceivable that the leadership styles change with fluctuations in degrees of emotional involvement and formal commitment during the career.

The mobility pattern and expectations regarding it have far-reaching significance for the style of leadership. The question can be analysed further by directly examining aspects of the succession process separately.

A person's style of leadership can be affected by his anticipation of promotion to another office. This anticipation, in-turn, can modify the behaviour of the promotable's own subordinates who are "next-in line" and develop new expectations because of it. The leader's behaviour seems to change with a cycle of promotion which includes the opportunity to be promoted, the promotion itself, and re-entrenchment afterwards. Levenson has observed for example that while striving to gain favourable ones he knows that he will be promoted, he again relaxes control, having less time or incentive to police subordinates as he learns his new job. At this same time, the leader also scrutinises his subordinates for a replacement which, in turn, undoubtedly constrains some of them. Moreover, in training someone to fill his position, the leader exposes much information about the work that was formerly secret.

Then, for a time following his promotion, the promotable is obliged to "prove" himself to the superiors. He may retighten control at this time, but once he has demonstrated himself, the leader may adopt still another style of leadership until there are prospects for another immediate promotion. Thus, cycles of authoritarian and democratic control, and other leadership characteristics which are often attributed to personality traits, are actually variables produced by the pressures of the situation itself.

The unpromotable supervisor probably has as significance effect on his subordinates' behaviour as the promotable. For when the supervisor is unpromotable, his subordinates' opportunities are restricts as well. Levenson outlines four possible reactions to unpromotable. First, the unpromotable one may withdraw, that is, leave either the organisation or the profession. However, whether this alternative is feasible depends on still other situational factors, such as age commitments and financial and social commitments. Or, he may region himself completely to the situation by abandoning ambition and other rewards, such as esteem and seniority. He may also adopt uncreative tactics designed to make supervisors notice him; he may "overcomes," or try to please indirect superiors, or become assertive at meeting. Finally, the unpromotable may rebel, and seek to displace his immediate superior, he may stage a showdown, seek to embarrass his supervisors every himself more cleverly, and so on.

The presence or absence of opportunity, then clearly administrates the influence that the situation has on behaviour. Aggressiveness withdrawal, and ritualism can be accounted for by the dead-end character of some jobs rather than personality traits. To the extent that the "dead-end" career is characteristic of the teaching profession, the about thesis is a promising point of departure which may have far-reaching significance for an understanding of the behaviour of teachers and their supervisors.

In periods of organisational crisis, leaders tend to assume note authority, to centralise power, and to demand greater conformity even subordinates. Conceivably, as organisational stress relaxes, decentralisation may increase (though not necessarily as quickly as centralisation develops from crisis), randa, and delays at promotion and contract time are well designed to imbue the organisation with an element of minor crisis.

A great deal can be learned about patterns of leadership by observing organisational tempos, the rise and decline of pressures generated by deadlines and by close supervision. For example, the school principal will sense that he has more authority, greater responsibility for the school, and more obedience from subordinates when the school is being "inspected" by the superintendent or visited by State Department of Education representatives or by a parent group. Similarly, teachers seem to be more "official" with students on the first day of school, and more informal on the last day. They also probably assert more officiousness at test time. When an important bond issue or consolidation vote is deciding the fate of the school, the administrator tends to become more directive.

In general, during crises the consideration role declines in favour of the initiating functions, and subordinates' wishes become less relevant to the concerns of the leadership. Halpin's typology of leaders based on their relative stress on initiation and structure needs modification to include the cyclical changes in these styles over a period of time as pressures ebb and flow. Theories of action provide the best means for confronting those organisational variables which are complicated by changes through time. We refer to natural system models such as that of Talcott Parsons.

Precisely because he is granted more authority in times of crises, the leader's position in most vulnerable then. To fail at a time of major crisis usually means the loss of the leadership position, for ineffective leadership is at no time more apparent nor more fateful. Consequently, the defeat of an important bond issue may result in a change of administration.

However, even the administrator who has won his cause — a bond issue, or a fight with the city council or the school board — may have jeopardised his position if he had to wield too much power or alienate too many persons in the process. The successful leader usually makes enemies in the process of a fight, or he may have become so committed to his supporters that the alternatives which remain open to him are so severely restricted by his obligations to them that his power is virtually destroyed. Often his best recourse after an all-out fight is to resign, whether he won or not.

"Busyness." An element of crisis, at least the appearance of it, is also used by some subordinates to strengthen their authority. "Busyness" is the art of appearing busy. It must be admitted as one of the effective ploys of "impression management." It conveys to observers the subordinate's assumed importance. Indeed, "busyness" is so effective that the casual observer cannot distinguish the busy from the productive members of the work group. However, in announcing his presumed importance the busy person is announcing his ambition to be important, which is significant in itself. It is this common element of "ambition" which links the unimportant and the important busy persons.

"Busyness" assumes characteristic levels of activity among various occupational groups. Most nurses, for example, complain that they are "too busy." Yet upon observing them at work in hospitals, it seems that the complaint is voiced more consistently than the actual workload would warrant. That is, even when she is not personally busy, a nurse will complain of it. Occupation-wide ideologies create exaggerated illusions of overwork. The statement that "most people in the occupation are busy" comes to mean that "I am busy." Characteristically, occupations striving for greater social status, including those in the process of professionalizatton, complain most about being busy. It is symbolic of dissatisfaction and ambition. Applied to teaching, the question is, to what extent do teachers' complaints about being overworked reflect dissatisfaction with their status. Without implying that most teachers do not work hard, it may be that complaints about hard work are indicative of a more fundamental condition of the occupation the drive for the status that comes with busyness.

The way a leader performs the leadership function depends to some extent on the risks he is both able and willing to endure. The leader's feelings about the security of this position vary independent of such personality traits as "confidence." His security is dependent on the permanency of his position and the guarantees that he will continue to occupy it. His personal commitment to the position is also important, of course. His position is important to him to the extent that there are few other acceptable jobs available to him. In short, his security is a function of the permanency of the position, his power to hold it, and the available alternatives.

During periods of proposed change his position is particularly vulnerable. A principal may resist efforts of the school board to bring in an "outsider" to fill the position for which he aspires. He may also resist the establishment of a position which has authority over his, thus decreasing the authority of his position. But he will not resist such a position if he expects to be promoted to it.

Similarly, for example, whether one reacts conservatively to proposed school district reorganisation plans depends not only on whether the plans threaten to displace him, but on his commitment to the job. It is precisely because volunteer groups such as the PAT are not fully committed to their position that they are in a position to take a liberal stand on proposed educational changes.

As suggested, the amount of commitment to a position depends upon the relative availability and importance of other positions.

In addition to personal commitment to his position, it was also suggested that the leader's behaviour depends on the security of the position itself. A person needs a certain amount of security to take a risk. So, it can be expected that the leaders who support educational changes hold positions in the largest school systems in the country which have the backing of the nation's most powerful economic and political leaders. This is somewhat contrary to the stereotyped notion that persons most entrenched in power are the most conservative defenders of the status quo. This may have some basis in fact, the upper class is both powerful and conservative – but historically the striving middle class has probably been more anxious and insecure about its status than the upper class.

Considering the organisation, the question of who is the most conservative leader depends on the relative balance between commitment to office and the security of the office. Where commitment is lowest and the office is most secure, leaders can afford to be less conservative. Conversely, conservatism is highest among leaders who are most committed to relatively insecure, unpowerful positions.

The size of an organisation is another facet of the leadership situation. From one study of 500 students who described the groups of which they had been a member, it was found that as group becomes larger, demands upon the leader's role become greater and more numerous, tolerance for leader-centred direction of group activities increases, and the group generally becomes more bureaucratic - that is, rules are enforced impartially, concern with administrative problems increases, and firmness towards subordinates increases.

These findings have several implications. First, they stress the fact that the problems of leadership become accentuated with bureaucratisation. The challenge of leadership in the future will increasingly be the problems generated by bureaucracy, and as these problems become more complex there is little prospect that they will be solved by "experience" or with mediocre training. Second, with his organisation increasing in size, it is difficult for the leader to give attention to the special problems of subordinates. It becomes more difficult to display the special leadership qualities of consideration; but perhaps consideration is also less demanded of the official leader as informal leaders assume this function through specialistic divisions of leadership functions. Finally, the evidence points to the dangers of projecting small group laboratory studies of leadership to complex organisational settings.

Interconnections

Two component of the leader's relationships to his subordinates are of special importance. On the one hand, the leader is expected by superiors to initiate ideas, maintain group norms, and act as final arbitrator of decisions. At the same time his relationships to subordinates concern "taking care of others," maintaining a humanitarian or, at least, an objective attitude, and in general not acting "like a big shot." When emphasis is placed on the consideration role, the leadership is sometimes loosely characterised as "democratic," while emphasis on initiating activities is often attributed to authoritarianism. Several studies have shown that skills in fulfilling these "initiation" and "consideration" roles are associated with good leadership ratings. One investigator concluded from a study of chairmen of liberal arts college departments that chairmen with good reputations as administrators are those who rate high on both the consideration and initiation dimensions.

A similar conclusion was reached from a study of eighty-nine aircraft commanders. Good r[illegible]gs of the commander by his superiors were associated with high initiation structure scores (originating new ideas or practices, maintaining standard operating procedures with his crew, giving regular and clear assignments to subordinates). Good ratings by his subordinates, however, were associated with high consideration scores (doing personal favours for crew members, being friendly, treating the subordinates as a social equal, looking out for crew welfare).

There are times when these two relationships require the leader to behave inconsistently. Telling others what to do and maintaining official standards may create resentment among subordinates, while establishing equal relationships with subordinates may prevent the leader's taking official action against them, and may jeopardise his relationship with his own superiors. A division of leadership labour is one solution to the dilemma. Some leaders seem to depend more on their initiation role, letting other members of the group take care of the needs and feelings of subordinates. This division creates at least two distinct types of leaders, those responsible for pursuing the group tasks and supporting its official structure, and those who are responsible for maintaining group sentiments, or its cohesiveness.

There is also reason to believe that the relative emphasis which an official leader gives to each of the elements is conditioned by his rank and his advancement opportunities. In one study of military leaders, it is reported that army officers, more than non-commissioned officers, competed openly with their peers and with their superiors, while they competed less with their own subordinates. On the other hand, although non-commissioned officers were more aggressive with their subordinates, they maintained less "social distance" from them than officers did. The lower ranking officials seemed to rely on the consideration leadership behaviour more than higher ranking officers did. The consideration function is perhaps left to informal mechanisms to a greater extent at high levels of command than lower ones. Similarly, initiation seems to be more informal at lower echelons.

Like so many other characteristics of the organisation, their initiation and consideration behaviours often necessitate compromise in practice. Styles of leadership are undoubtedly associated with the nature of these compromises. Halpin has developed a typology of leadership styles in terms of four combinations of initiation and consideration structure. School superintendents who were high on both dimensions were assumed to be the most effective; only eleven of fifty superintendents were placed in that category by both their staffs and the school board.

Division of Society

The initiation and consideration roles are no different that subordinates and superiors, leaders and followers, are characteristically separated by formal and informal barriers. Of this, one school superintendent has said, "You can become too – maybe too putting yourself on too much of an equal – to be too friendly. Now there has been some little thought that has come to my mind that maybe I have been a little too friendly. Some of the superintendents in a Massachusetts study followed a policy of having no intimate friends in the community in which they worked, and they were reluctant to develop friendships with their staff member. The social distance that characterises leader and subordinate relationships is most apparent in the military where there are official restrictions against fraternisation, and separate social, eating, and sleeping places (they even wear their status on sleeves and shoulders).

In maintaining social distance from subordinates, the leader is handicapped in understanding them and their wishes, which generally prevents the most effective fulfilment of the consideration role. This is a structural source of ignorance which may leave the needs of subordinates unanswered. This social distance is one reason that informal leaders arise within the subordinate ranks to take care of the consideration roles neglected by distant superiors.

If it handicaps the consideration role, then why is social distance so characteristic of leadership positions? From an examination of this question, one investigator concludes that it is not because subordinates' feelings of familiarity with their leader hinder their own effectiveness; nor is it because the knowledge which subordinates have of their leader's defects reduces their confidence in him. Rather, social distance is designed to protect the leaders' performance of initiating behaviours. Barriers protect the leader from any involvement with subordinates which might influence his ability to reach decisions that contribute to the goals. As one school superintendent put it, I mean they would feel, 'He's an old buddy of mine.' If you had to crack the whip a little bit, or set down some rules, then they'd be offended quicker. Social distance seems to produce an atmosphere favourable to rationality by preventing unnecessary personal commitments from interfering with the leadership role.

This is not to say that social distance is without benefit to subordinates, for it does benefit them. At the same time that he is aloof to his subordinates, the leader is associating with other, perhaps more influential people – that is, school and community leaders. This puts him in a position to make contacts with those who can achieve salary increases and otherwise protect the interests of his subordinates, advantages which they could not have achieved themselves.

One researcher suggests that social distance is partly explained by virtue of the fact that the satisfactions realised through peer relationships might be jeopardised by too close relationships with superiors.

It is relatively more difficult for leaders in some positions than those in others to maintain social distance. Lower ranking officials are often in closer physical proximity the subordinates than higher officials are to theirs. This may help to account for relatively greater emphasis (noted earlier) that lower officers give to consideration. Teachers, for example, having no office of their own, cannot completely withdraw from students as principals can withdraw from teachers. They are under pressures to be popular and a "good Joe," with no intermediary to bear the brunt of resentment; and not having particularly high social status, as well as being subject to the ridicule of their students and parents, teachers are especially likely to rely on their consideration role, which makes them vulnerable to the charge of "favouritism." There is always a threat that the teacher will succumb to humanistic conceptions of students rather than to official ones, that they will like some of their pupils and dislike others, because of their close special relationship.

Nature of "Subordinates:" The concept of leadership is meaningless except as it implies a set of relationships. It cannot be understood without understanding the other positions with which leaders must deal. Of particular importance is the nature of the subordinate position.

The practice of speaking about subordinates categorically obscures their characteristic diversity. Subordinates do not represent a unified front, but rather a mixture of subgroups which support or resist different leaders in various degrees. In a study of sociometric choices in a training school for girls, for example, it was reported that there was very little overlap among individuals who supported different group leaders. Large organisations, particularly, are characterised by heterogeneity and dissent among subgroups who are divided among themselves on the basis of their allegiance to different leaders.

This heterogeneous character of subordinates influences the structure of leadership. An array of leaders develop who are supported by various factions of subordinates. A portion of the power of the leader is dependent on the relative power of his supporting (action compared with the elements that oppose him.

School and Community

School: a Community Centre

The present conception of education differs radically from the old conception of education and consequently the functions of the present school differ from the traditional school. The traditional conception of a school has been too academic. According to it the school was "a knowledge shop and the teachers information mongers." Teaching activities were carried on in it in an academic seclusion out of touch with the social and economic life surging around. It was preoccupied with one narrow but clear-cut problem – how to prepare its pupils to pass certain examinations which were prescribed by the educational authorities. Knowledge was the be-all and the end-all of education. The modern sociological view of education postulates that "the school should constantly draw upon social life and activities for its subject-matter, its methods of teaching and its methods of work. There must be a conscious and continuous intercourse, a free give and take, between the little world of the school and the bigger one outside."

Secondly, the school is not a place where merely the children are educated but the whole community. With tons of money invested in school buildings and equipment, it is a poor economics as well as bad educational philosophy to restrict its use to a few hours each day, five and a half days a week and for only a few months in the year. Its use must be made of by the whole community regardless of creed or religious or political affiliation. The Folk High Schools of Denmark, the Tushegee Institute for Negros in the United States and the Gary Schools of USA furnish notable instances of the community service which schools can render. The school building and equipment must be thrown open for public use after the regular school hours are over.

Thirdly, the tax-papers will feel that they are really getting a fair return for their money and will take all the more interest in the school if it is taken out of its isolation.

Fourthly, the school teachers will get chances to become the leaders of the social group and can improve their position and prestige.

Rural School and its Community Functions: What has been said above applies to all schools, rural and urban. But India being a village country, the rural schools have an important duty to the nation. Rural uplift is the only solution if we want to see India standing in the row of all advanced countries. The Community Projects and National Extension Service Schemes have in view the object of making our villages pulsate with life as in the old days and not continue to decay as at present. Gandhiji raised the slogan "Back to villages". The village schools have to play an important role in the reconstruction of the villages. Mr. F.L. Brayne remarks that there is no better and cheaper agency possible for the remaking of village India than the village school.

The village school must serve as a centre of light to the villagers and supply cohesion, initiative and knowledge required for the great task of rural reconstruction. To quote Mobel Carney, "The interpretation of the function of the country school as one of social leadership and community service implies that it must reach not only children but adults also. No argument is needed to establish desirability of this service. To educate properly is to cause a change of conduct and the hypothesis that the country school shall so govern its instruction as to cause a change in social attitudes and community conditions and relationships in no way clashes with its original purpose (to educate)."

Village schools are to be organised on a different pattern than that of an urban school. At present the position is that there is no particular difference between a rural school and an urban one. So far as the aims, curricula and methods of teaching are concerned, both are alike. The same curriculum and the same methods cannot be suitable to rural and urban children. So the present curriculum should be 'ruralised.'

Community Serving Methods

The school will employ the following three methods to influence the community:

1. Producing good individuals or imparting training for citizenship.
2. Organising community service programmes.
3. Throwing open the school building for social, cultural and recreational programmes.

Education for Citizenship

Vocational Efficiency: The old system fails to prepare students for a vocation and simply produces persons who are fit for white collar jobs only. Such individuals fail to harness the resources of the community and are unable to serve the community to which they belong. In a Basic school adequate stress is laid on the development of sound working attitudes and habits and appreciation of the dignity of labour. The introduction of various crafts helps the students to develop vocational efficiency and thereby saves them from becoming misfits in life.

Appreciation of the Rights and Responsibilities: A Basic school provides many opportunities to the students to appreciate that rights and responsibilities go hand in hand. The introduction of student self-government in a limited form helps very much to teach this lesson to the students.

Physical and Health Development: It must be constantly impressed upon children that a healthy body contributes much to become happy and efficient citizen. A course in scientific knowledge of food and nutrition should form a part of the curriculum. The importance of good food and living in healthy surroundings should be told to them. School building, school hostel and school surroundings should be made model of cleanliness. The idea is to develop good health habits in the students so that they may spread these in the locality to which they belong.

Appreciation of the Inter-relatedness of Different Agencies: A man is never self-sufficient and is dependant upon others for the satisfaction of his needs. From the very beginning a child should be made to realise that cooperation is the basis of all the activities of man – social, economic, political and educational.

Organising Social Service Programmes

The field of social service activities is very vast and may be grouped under the following heads:

Physical Activities: The students may be asked to clean the lanes and homes of the locality, to construct open air theatre, to dry up marshy lands, etc. These activities have immense educational and social values.

Cultural Activities: India is a land of festivals. The school should organise seasonal, national and cultural festivals and invite parents to attend these. The elders or the experienced persons of the locality may be asked to address the audience.

Campaigns Against Social Evils: Periodical campaigns by the school may be organised and the evils of drinking, gambling and borrowing pointed out.

Literacy Campaigns: Night classes for the adults may be started and conducted by the elder students of the school and the teachers also.

Miscellaneous Activities: Social service by the students may be rendered during fairs, floods, epidemics, etc. Occasional trips to the various parts of the locality may be arranged by the school authorities, with the idea of making the students know of the problems confronting the community and creating an urge in them to remove defects which they notice. Local surveys may take the form of projects like the extent of literacy in the locality, cleanliness, living conditions in the homes and the like. The school may serve as a 'clearing house' as K.G. Saiyidain points it, where teachers and parents of the boys meet and discuss in a friendly way their problems, social, economic and educational, with which they are faced and in which both are interested.

Utilising School Building: It will be very desirable if the above mentioned programmes are organised in the school premises. Besides in places where there is no public library, the school should consider the possibility of throwing the school library open to the public after school hours. A special section catering to the needs of the adults may be opened.

Amendment Projects

1. Construction of school buildings; provision of additional accommodation.
2. Electrification of buildings.
3. Construction of quarters for teachers.
4. White-washing and repairs of buildings.
5. Construction of compound walls, provision of sheds for cooking free meals and provision of fence for garden.
6. Donations of lands for free meal scheme.
7. Donations of lands for agricultural and garden purposes.
8. Acquisition of land for playground.
9. Provision of classroom equipment like screens, painting of black-boards, supply of classroom furniture, supply of equipment and teaching aids and supply of portraits of great leaders.
10. Supply of slates and textbooks to poor children.
11. Supply of free clothing.
12. Supply of personal hygiene materials (soap, towels, mirrors, etc.)
13. Supply of grain, firewood, etc. for free meal scheme, supply of cooking vessels for free meal scheme and supply of plates and tumblers, etc.
14. Provision of drinking water.
15. Provision for night study and supply of petromax and hurricane lights.
16. Supply of equipment for handicrafts like sewing machine, etc.
17. Supply of musical instruments and radio.

18. Supply of play and hobby materials.
19. Supply of garden tools.
20. Maintenance of existing wells and sinking of new ones.
21. Provision of sanitary conveniences.
22. Supply of books, newspapers and magazines.
23. Provision of first-aid materials.

The Values

First, the work of planning and organisation of these programmes should be undertaken very carefully beforehand by the teachers and students in collaboration.

Secondly, programmes should be so varied that they give scope for many students to participate in them.

Thirdly, cooperative aspect of the work should constantly be emphasised.

Fourthly, these programmes should be linked up with the academic work also.

Fifthly, the quality of the work done by the students should improve gradually.

Lastly, the students must be made to feel that year after year they are not merely carrying out a routine activity, but they are doing something which is really useful to the community and that their abilities are utilised for noble causes.

Community and the School: "What we would like to see is a two way traffic," thus wrote the Secondary Education Commission. In this context they have suggested that interested members of the community, engaged in various useful vocations and professions should be invited to the school from time to time to talk about their particular work, to show its place and significance in the life of the community.

Parent-Teacher Collaboration

In the words of S. Balakrishna Joshi, "Close cooperation between the parent who is the first teacher and the teacher, who is the second parent, is the very foundation on which rests the fruitfulness of the training imparted in our institutions." Mr. George Tomlinson has fully stressed the need of cooperation between home and school in these words: "Let us fashion our homes and organise our schools with the well-being of children always in mind. In particular, remember that any clash between parents and teachers must always be harmful to the child. Harmonious working together can alone bring us the results we want." Teachers are interested in the welfare of the children, they want their pupils develop mentally, physically, morally and socially. So is the case with the parents. Hence the need for united efforts on the part of parents and teachers who are the custodians of the welfare of children.

Significance of Parents' Day

Importance of the Parents 'Day from the Teachers' Viewpoint

1. They get an opportunity to acquaint the parents with the various activities of the school.
2. They get an opportunity to explain to the parents what future plans the school has for the welfare of the students.
3. They get an opportunity to request the parents to render assistance to the school.
4. They get an opportunity to exchange views.

Parents' Viewpoint

1. It provides them an excellent opportunity to see the working of the school.
2. It provides them an opportunity to know what their children are doing at school.
3. It enables them to discuss certain specific problems of their children with the teachers.

From Pupils' Viewpoint

1. They feel immense joy when they see their parents visiting the school.
2. They try to put in their best for making the programme a success. Esprit de Corps is developed in them and they feel love for the school.

The Curriculum

It should be a full working day. The parents should be taken round the school so that they may see how the classes are held and where their children sit. They should get an occasion to see the actual teaching done in the school.

Different co-curricular activities such as debates, dramatics, etc. may be organised.

Children's work like calligraphy, models, maps, charts, etc. may be exhibited in one of the rooms of the schools.

The parents may be taken to special rooms, e.g. Geography, History, Science, Craft, etc.

School sports may be organised on this day or it may take the form of prize distribution or School Annual Day.

Adequate preparations should be made so as to give a good show of school activities. A light refreshment to the parents may be served if the funds allow.

The Requirement

From the Point of View of Teachers

1. To seek maximum information from the parents about the child to have fuller understanding of him.
2. For getting help from the parents in the regular attendance of the child.
3. For getting help in regular supervision of home work of the child.
4. For providing the child necessary equipment at proper time.
5. For getting help in stressing upon the child the necessity of having respect for the school laws.
6. For securing funds and gifts for the school.

From Parents' Point of View

1. They need information regarding the progress of their children.
2. They want necessary warning regarding expenditure on books, etc.

3. They need information about problem situations before they become serious.
4. They want careful assignments.
5. They need advice as to how they could supplement the efforts of the school in promoting the progress of their children.

The Problems

From the Side of the Teachers

1. Their unwillingness to face criticism.
2. Their conservative outlook and doubts about the success of this programme.
3. Lack of time to make sincere efforts in this direction.

From the Side of the Parents

1. Indifference of the parents to school work.
2. Busy life of the parents and lack of time to contact teachers.
3. Ignorance on the part of the parents that they can render any assistance to the child in his work.
4. Their suspicious nature regarding the motives of school authorities.

Securing Cooperation Methods

In view of the mass illiteracy in our country, the initiative for the home-school cooperation rests upon the teachers. To quote S. Balakrishna Joshi, "Parental cooperation cannot be had by learned discourses on the subject or by periodical pamphlets posted to the parties. It cannot be created overnight. We cannot wake up one fine morning and find the parents pitch forked into cooperation. It is based on a series of silent acts of sincere service to children. It is the natural consumption of devoted deeds of dedication at the shrine of Duty." All possible avenues should be explored to establish contacts with the parents. Following are some of the main methods of securing cooperation of the parents and to make them interested in school work.

Good Treatment: Whenever the parents visit the school they should be shown proper courtesy. Patient hearing should be given to their complaints.

Personal Contacts: The parents of those children who do not show good results in monthly tests or terminal examinations, may be requested to visit the school. Positive suggestions regarding the future progress of the children should be given to them. Frank and free talks would result in active cooperation.

School Reports: These include:

Casual Letters: These should be used only when it is not possible to have personal contact with the parents. The parents of those students who reside in the hostel may have to be contacted through letters.

Progress Reports: These are sent to the parents from time to time. It is a useful device to keep in touch with the parents. Progress report should be very comprehensive, touching all the different aspects of the life of the child – mental, moral, physical, academic and aesthetic, etc.

Participation by the Parents in School Programmes: 1. Some parents can lend a helping hand in raising funds for the school. Their help should be secured. This will develop in them a sense of belongingness to the school. There are many other ways through which the parents can help the school authorities, e.g., donating or lending books, giving stipends to the poor students, arranging midday meals for the poor, etc. The school authorities should not hesitate to utilise the knowledge, skill and experience of the parents in supplementing class teaching. A doctor may be requested to give lectures on Hygiene and Physiology. Similarly, the services of the parents in other fields may be utilised.

Home Visits by the Teachers: These should always be friendly. The false notions of prestige should not stand in the way.

Association of the Parents in the Management of the School: Representation should be given to the parents on the board's management. But in some cases this may prove very harmful to the efficient working of the school.

Making the School as a Centre of Community Service: National, social and cultural functions should be organised in the school. Labour weeks may be organised wherein students render some social service to the community.

Formation of Parent-Teacher Association: The meetings of the association provide an opportunity to the parents to make suggestions for the improvement of the school. The head and the members of the staff acquaint the parents with the current thought and practices of education. The sphere of the working of such associations should be clearly defined so as to avoid unnecessary interference on the part of the parents in the internal organisation of the school. It would be better if the parents of the students of one class are invited to meet the teachers of that class, at one time. The smaller the group, the better would be the results.

In the end we may again stress the importance of making the parents interested in school work. "The ordinary machinery for good school life," writes Thring, "is a thing of calculation and measurement but all discussion is in vain unless parents are interested."

Glorified Essence

"A 'people's school' must obviously be based on the people's needs and problems. Its curriculum should be an epitome of their life. Its methods of work must approximate to theirs. It should reflect all that is significant and characteristics in the life of the community in its natural setting," observed Prof. K.G. Saiyidain.

The importance of linking the school life with the life of the community has been stressed by he Secondary Education Commission in these words, "The starting point of educational reform must be the relinking of the school to life and restoring of the intimate relationship between them which has broken down with the development of the formal tradition of education." Dewey also in his book, School and Society emphasises this when he writes, "The great thing to keep in mind, then regarding the introduction into the school of various forms of active occupation, is that through them the entire spirit of the school is renewed, it has a chance to affiliate itself with life to become the child's habitat, where he learns through direct living, instead of being only a place to learn lessons having an abstract and remote reference of some possible living to be done in future. It gets a chance to be a miniature community, an embryonic society. This is the fundamental fact and from this arise continuous and orderly streams of instruction." According to Raymont, "The school should be a bit of life, or rather, perhaps, it should be, in some sort a miniature of the best elements in the great society outside."

K.G. Saiyidain describes the significance of school – community relations in these words, "If we continue the present practice of working in isolated compartments, the school not enriching the community, the community not supporting the school, not only will it defeat our real educational objectives, but whatever education we provide, will be stale and anaemic."

Ryburn describes the relationship between the school and the community in these words, "There must be a vital connection between the life of the pupils in school and the life of the community from which they come. There must be a vital connection between the school, which is the corporate life of pupils and teachers and the community. Otherwise, the school can never succeed in its aim of enabling its pupils to go out and to face society and make necessary adjustments nor can it, as a corporate body, even have the vital influence on the community which it ought to have."

School is a social institution set up by the society to serve its ends. It is charged by the society with the duty of training and bringing up the students so that they may be able to take part effectively, harmoniously and efficiently in the group to which they belong. They cannot inherit automatically their social heritage as they inherit their ancestral property. Constant efforts have got to be made to bring the activities of the school closely in touch with the life outside the school. School is a society in 'miniature'. It must be brought out of its isolation and connection with all the nobler, and the worthier aspects of the life of the community. Knowledge gained in the school must be linked with the life of the community. Failure to do so, will only produce shiftless, aimless, helpless and perhaps worthless individuals. All the possibilities of connecting the school life with the community life be explored and the child fully prepared to take his rightful place in the community. Curriculum, methods of teaching, school building, library and the experience of the teachers are the various tools to link the school with the community.

Significance of Society

Social justice, as an objective of state policy, has been emphasised in the Indian Constitution. The preamble to the Constitution states

> "We, the people of INDIA, having solemnly resolved to constitute India into a Sovereign Socialist Secular Democratic Republic and to secure to all its citizens justice- social, economic and political;_______"

In the list of 84 values appended to the questionnaire social justice was mentioned as one of the values. The following statement, having a bearing on this value, was included in the questionnaire for the general population and the people were asked to respond to the statement:

" In Indian Society, social justice is denied to women". (G-S)

Social Equality

Sl. No.	*Category of Respondents*	*Number Agreeing*	*Number Undecided*	*Number Disagreeing*
1	2	3	4	5
1.	Total population of respondents. (881)	586 (66.5)	106 (12.0)	189 (21.5)

Contd..

1	2	3	4	5
2.	Male respondents (671)	395 (64.0)	65 (10.5)	157 (25.5)
3.	Female respondents (264)	190 (72)	41 (15.5)	33 (12.5)

It is, thus, seen that 72 per cent of female respondents felt that social justice was denied to them whereas 64 per cent of the male respondents subscribed to this view.

Human physiology and culture play an important part in determining the role of men or women in the group. Biologically speaking, man is larger in size and also sturdier than woman. Besides, woman has to bear and nurse children, which are bound to keep her indoors more than men. These factors determine different kinds of role for men and women and, to some extent, bestow dominant position to males.

The Vedic people accorded a respectable status to woman. The Vedas and the Puranas picture the female as divinity. She is worshipped as Adi Shakti (First creator), in the form of Mother Parvati and destroyer in the form of Katika. Many great Grama Devatas (village gods) of south India – Mariamma, Elamma, Gangamma, Bhadrakali, Pida – are women. The names of the seven magic drugs and vegetables given in the Agni Purana and which are to be placed within amulets for ensuring victory over the enemies, are female – Maahakali, Chandi, Varahi, Isvari, Indrani, Vala and Gauri.

Indian women, in ancient days, did not lead a secluded life which circumstances forced on them in the later period. Women had the privilege of receiving high education. The custom of Upanayana of girls prevailed down to the Sutra period. Several Vedic hymns were composed by female Rishis. Some of the women, who composed Vedic mantras 'are Apata, Ghost, Hopmudra, Indrani and Sarswati still have a great influence on the Hindu society. Women like Sita, Ahilya, Draupadi, Tara, and Mandodri are respected by both men and women alike even today. At a meeting of theologians convened by the scholar-king Janaka of

Mithila, Gargi carried on a lively discussion with the great saint Yajnavalkya till she had extracted the definition of Brahman from him. Maitreyi, the wife of Yajnavalkya, was also well-known for her scholarship.

During the famous theological debate between Adi Shankaracharya and Mandan Misra on the interpretation of the Vedas and the Upanisads, in which the former came out victorious, it was Bharati, the wife of Mandan Misra, who acted as the judge. She had been educated at a gurukul in Prayag and she was known for her scholarship. At a later stage, Bharati herself had a debate with Adi Shankaracharya who was considered as one of the greatest masterminds of the Hindu philosophy.

The importance of female education is shown by the fact that according to the Atharva Veda, an educated brahmcharini had better prospects for marriage than one who was uneducated. Girls were then not married at an early age. In a hymn in the Rigveda (X, 85, 21 22), the god of marriage, Visvavasu, is asked to go to a maiden who has attained the sign of marriage, whose body is developed and unite her to a suitable husband. The well-known ancient practice of Svayamvara indicates that women had a say in selecting their life partners. Widow-remarriage was allowed. A hymn in Rigveda (X, 18.8) says, "Rise up woman, thou art lying by one whose life is gone. Come to the world - and become the wife of one who is willing to marry thee." In religious ceremonies, wife had to be always by the side of her husband.

All this indicates the high position that women enjoyed in the Vedic India, though there are some contradictory assessments about the place of women in the society in the subsequent periods. Yajnavalkya and Vatsyayana give a positive assessment of the female character, whereas Dharma Sutras, Dharma Sastras and the *Mahabharata,* at places, adopt uncomplimentary view of women In *Ramcharitmanas,* the great poet Tulsidas says that a drum, an uncultured person, an animal and a woman deserve to be beaten. The great Codifier of the Hindu laws, Manu, says, "The gods reside where women are respected" He also says, "In childhood a woman must be subject to her father, in youth to her husband, and in old age to her son; a woman must never be independent because she is innately as impure as falsehood, the Lord created woman as one who is full of sensuality, wrath, dishonesty, malice and bad conduct". Vasistha says, "A teacher (acharya) is ten times more venerable than a sub- (upadhyaya), the father a hundred times more than the teacher, and a mother thousand times more than the father":

During the medieval period of Indian history, a considerable, change took place in the status of women. "Widow remarriage was forbidden. The position of women was not as high as in ancient India – In a word, her (woman's) life was in a state of perpetual wardship – the Shastras laid down that the woman was socially and spiritually inferior to man. The position of women worsened during the period of the Sultanate of Delhi, The institution of the devadasis, in principal temples, was then in existence. Al-Beruni says that the priests were opposed to the institution of devadasis (beautiful maidens in temples practising devotional music and dance), but the Kings, for the sake of revenue, maintained them even when it had degenerated in character."

Necessary Provisions

Articles 15(1) and 29 (2) of the Constitution of India, reproduced below, originally did not stipulate any discrimination between various citizens.

Article 29 (2)

"No citizen shall be denied admission into any educational institution maintained by the State or receiving aid out of State funds on grounds only of religion, race, caste, language or any of them."

Article 15 (1)

"The State shall not discriminate against any citizen on grounds of religion, race, caste, sex, place of birth or any of them."

Article 15 (3)

"Nothing in this article shall prevent the State from making any special provision for women and children."

However, the following addition was made in Article 15 of the Constitution (First Amendment) Act, 1951:

Article 15 (4)

"Nothing in this Article (i.e. Article 15) or Article 29 shall prevent the State from making any special provision for the advancement of any socially and educationally backward classes of citizens or for the Scheduled Castes and the Scheduled Tribes.

There is no denying the fact that for a long time, low castes in the Hindu society were given a very harsh and a raw deal. For example, in southern Travancore, one mark of respect to the superiors was a bare breast. The priests honoured the deities in this way. By the same token, lower castes women were not allowed to cover their breasts. It was only in 1829 that the government of Travancore allowed the lower -caste women to cover their breasts, but not in the manner of higher caste Hindu women. Higher caste and lower caste Hindus used different language; as for example different words for water. Such things are, now vanishing.

Rudolphs states that the following factors are gradually obliterating the caste differences in India

"Four processes are making Indians more alike, and, in doing so, are laying the necessary but not sufficient conditions for national integration; ascriptive boundaries are expanding; the culture and the status of the twice-born varnas are spreading to Sudra castes; Westernisation is effecting the ideas and occupations of broader sections of society; and secularism is dismantling ritual barriers and disarming sacred sanctions.

Due to progressive legislative measures, expanding educational facilities, gradual improvement in the economic status of Scheduled Castes and the Scheduled Tribes, people belonging to the so-called higher castes and lower castes tend to meet on equal terms. Also on account of the mobility of many people from villages to cities, where the caste origins of people are unknown, caste considerations are getting diffused.

The Paradox: Paradoxically, certain Scheduled Castes now strive to retain their identity and would not allow higher castes to join them. "For example, in 1956 the dalits of Rajasthan were eager to disassociate themselves from their customary trade of leather working. By 1963, however, because of generous government financial and technical support to an industry capable of earning desperately needed foreign exchange and because of the advent of appreciable technical upgrading in the work involved, dalit were straining to exclude 'opportunists' from invading their ancient monoply."

Curricular Content: On the issue of social justice to women and SCs and STs, the following units maybe considered for inclusion in the curricula for schools and colleges. Preparation of exhaustive curricula for different age groups suiting their mental age is left to the specialists.

1. Broad picture of the present status of women in various sections of the Indian society and existing legislation on the subject.
2. Status of women in ancient and medieval India.
3. Female infanticide, malnutrition of females, rapes, dowry deaths, overburdening girls with household jobs, etc.
4. Emancipation of women in India and abroad.
5. Developmental programmes which can ameliorate the status of women.
6. The story of caste-system in India and how the system has been abused by the society.
7. Social and economic injustice meted out to the SCs and STs.
8. Steps so far taken to improve the lot of deprived sections of society. Steps taken in other countries to improve the lot of people belonging to backward sections in their respective lands.

It hardly needs to be emphasised that every member of society should get economic, political and social justice, irrespective of class, caste or sex considerations.

Violation of Social Norms: Nature observes its laws meticulously. Being a product of nature, man should, logically speaking, follow the codes of behaviour laid down by society. However, this is not always so. It would, therefore, be interesting to study why man violates the social norms and takes to certain behaviour patterns which are sometimes harmful to him, sometimes to the group around him and at times to both.

Anti-social Behaviour: A statement was incorporated in the questionnaire to assess the relative importance of certain causes which are considered to be at the root of an individual's or a group's anti-social behaviour. The statement was worded as follows:

"Anti-social acts originate from the following causes –

1. Financial hardships;
2. Lack of education among those who commit such acts;
3. Bad influence of mass media such as films, TV". (G-15)

Respondents were requested to grade the above causes. Weighted scores for the above causes are indicated below.

Wrong Actions

Statement	*No. of persons who allotted first, second and third position to this statement*			*No, of person, who expressed no opinion.*	*Weighted scores*
	I-Position	*II-Position*	*III-Position*		
(a)	362	259	145	115	66.296
(b)	292	269	206	114	61.390
(c)	109	238	417	117	46.2 per cent

Anti-social acts are of two kinds

1. Crimes where violence is used;
2. Crimes, which largely require the application of one's grey matter rather than physical force.

With the spread of education and the development of human brain, new types of crimes are emerging which require more of intelligence than force. As it is largely the white-collar class which is involved in these crimes, such crimes are known as white-collar crimes. Examples of such crimes are fabrication of fake university diplomas and degrees, immoral traffic, cheating, forgery, pilfering civic property, tax-evasion, distorted interpretation of rules and laws, smuggling, foreign exchange rackets, acceptance of kickbacks, etc.

There are many causes of the crimes. Of the three causes listed in the questionnaire, the respondents graded them in the descending order as given in the foregoing statement (G-15).

A study done in this connection regarding responsibility of the school for children's delinquency led to the following interesting and surprising results.

"After following a group of school children for eleven years, a team of researchers concluded that school relationships are the third most important contributors to delinquency, following only the family and the poor group."

Another study by Delbert S. Eliot found the school itself and not the family or "the streets" to be critical for generation of delinquent behaviour.

"Researchers have held that schools promote delinquency in the following way. Students who are unable to meet the expectations of their teachers are rejected by them. In this process of alteration, they become truant and rebellious and drift into association with the delinquent subculture. Eliot and Voss, by following 2600 junior high school students for four years, found that delinquency rate was significantly lower among those who dropped out than it was among those with similar academic records who remained in school."

Thus, the school teachers become responsible for delinquency of children in two ways:

1. By alienating students by adopting an attitude of indifference to those who do not perform well in academics;
2. By ignoring those who are dropouts due to one reason or the other.

In other words, in terms of the studies quoted above, indifference to education comes to occupy very near, if not actually, the first position, among the list of causes for committing crimes.

The following statement was included in the questionnaire for students to know their reaction to bribery, which is one of the white-collar crimes.

"It is pardonable, if a needy person accepted bribe". (S-3)

The following score-chart gives indication of the students views in this connection.

Permissiveness for Accepting Bribe

S. No.	*Category of respondents*	*Number agreeing*	*Number undecided*	*Number disagreeing*	*Number of persons who expressed no opinion*
1.	Total students population (267)	36 (13.5)	48 (18)	181 (67.8)	2 (0.7)
2.	School students (M) (70)	6 (8.6)	16 (22.8)	48 (68.6)	–
3.	School students (F) (86)	10 (11.6)	17 (19.8)	59 (68.6)	–
4.	College students (M) (34)	10 (29.5)	4 (11.7)	18 (52.9)	2 (5.9)
5.	College students (F) (70)	9 (12.9)	11 (15.7)	50 (71.4)	– –
6.	African students (M) (7)	1 (14.3)	– –	6 (85.7)	– –

As noticed above, comparatively the least opposition to the statement was from college-students (male). All other categories of students opposed the practice of acceptance of bribe to a greater extent even by a needy person. It can not be denied that indulging in bribes has greater bearing in one's attitude of mind than on one's needs. Once we fall a prey to this practice, the habit overtakes us and thereafter bribe is accepted not for meeting our needs, but to satisfy our lust for money. It is unfortunate that bribery has become rampant in India in recent decades.

An Indian politician once said, "If you strike a spade in Saudi Arabia, you well get oil, but if you do so in India, you will find corruption. In these circumstances, if one condones bribery, existing moral chaos in the society is bound to worsen. Therefore, the evil is rightly punishable under the law."

The statement included in the questionnaire on this issue was as follows:

"Salamat and John were discussing drinking. Salamat argued that one should totally keep away from alcoholic drinks. John, on the other hand, felt that drinking to a limited degree to keep company should be acceptable to the society.

Who do you agree with ?

1. Salamat, who totally rejects drinking

   Or

2. John who accepts drinking in moderation for company sake". (G-3)

Response-pattern to the above statement was as follows:

Abstinence Vs. Moderate Drinking

Option	*No. of persons agreeing*	*No of persons disagreeing*	*No of persons un decided or who expressed no opinion*
(a)	542 (61.5)	5 (0.6)	25
OR			
(b)	307 (34.8)	2 (0.2)	(2.9)

Total number of respondents was 881. Thus there was an above-average support for the option favouring total rejection of drinking.

While debating the above statement, one has to take into account several dimensions of this issue.

1. What should be taken as norm of moderate drinking?
2. If drinking is to be allowed, what should be the minimum age of the person before he takes to drinking ?

   When Mikhail Gorbachov came to power in the USSR in March 1985, in order to curb the drinking habit among the youth, he raised the drinking age in the country to 21 years.
3. There may be certain areas in which drinking has come to be connected with one's work. For example, business parleys and diplomatic jobs are considered to be more conducive to initiate a person into drinking habit. If one is to talk of total abstinence, one also must take situations like these into account.

   Though solitary or exceptional cases do not establish any rule, yet a concrete example here would provide a food for thought. A Minister of Information and Broadcasting in the Indian Union Cabinet in the seventies of this century did not drink. He was advised by his friends to start drinking, at least in moderation, otherwise, the journalist would not like to 'spoil their evenings' with him merely over a cup of tea. As incharge of Information and Broadcasting Ministry, the Ministers' frequent meetings with the journalists were highly essential. Notwithstanding the fact that he was keen on developing a proper equation with editors and correspondents of the newspapers as well as the writers' world, he did not take recourse to drinking, still the journalists and writers were his regular visitors, because he was highly intelligent and had enough to offer to them by way of news and have a purposeful dialogue with them on various political, economic and social issues. Later, when this very minister was appointed, in the Cabinet rank, as ambassador to the USSR, even in his diplomatic assignment he did not initiate himself into the drinking habit, yet he was highly successful in his mission. This example shows that in political and diplomatic jobs one can manage without drinking if one is properly equipped otherwise for the functions assigned to him or her.

Similarly in the business sphere, one can achieve the desired success by being resourceful and honest in dealings. Moreover, consuming a peg and offering one to a guest are two different things. If your work or the situation demands that you are to offer a glass of wine to a guest, it does not mean that you cannot get yourself excused from sharing the same.

4. There are people, who hold that drinking is outside the domain of social morality and it is purely a personal matter. As such, drinking should have no social stigma attached to it. Is it an issue of personal freedom or that of personal eccentricity? It has to be seen in proper perspective and the social hazards that drinking poses have to be taken fully into account.

   Further, it should be remembered that drinking is quite common among those persons also who later become drug-addicts or take to illegal distillation and illegal trading in spirits. Drug addiction and drug trading are assuming frightening dimensions in India. India was having more than one and a half crores of drug-addicts in the year 1988. Geographically, our country is quite vulnerable to drug smuggling and drug trading because countries located in the vicinity of India trade in narcotics. These countries are Nepal, Burma, Thailand, Pakistan and Afghanistan. Therefore, the question of drinking has to be viewed in this overall perspective and decided whether it constitutes or does not constitute a part of social morality.

The opposition to drinking is believed to be on the following grounds.

1. Alcohol (when taken in excess) depresses the activity of the central nervous system and thereby interferes with coordination, reaction time and reasoning ability. The harmful effects of excessive drinking on the liver are well-known; the end result may be cirrhosis, a condition in which liver cells are destroyed by alcohol and replaced by scar tissues. In addition to physical effects, alcohol contributes to several social problems. These social problems include suicides, murders, sexual assaults, release of animal instinct in man, broken homes, financial problems and numerous other anti-social and unhealthy situations. According to an assessment by the Soviet Union, one third of crimes in that country were linked to drunkenness.
2. Innumerable tests of physical and mental performance show that alcohol reduces visual acuity and delays recovery from visual dazzle, impairs taste, smell and hearing, muscular coordination and steadiness, and prolongs reaction time. It leads to alcohol dependence
3. The effects of alcohol and psychotropic drugs on motor driving have been intensively studied. There is impressive evidence to demonstrate that alcohol plays a huge part in causing motor accidents, being a factor perhaps in as many as 50 per cent. It is revealed that alcohol brings disaster on the road less because of lack of skill than because of defective judgement in relation to skill. For this reason, compulsory

roadside breath-test and provision for bloodtest (if necessary) have been acknowledged to be in public interest.

One may say that the issue under consideration was a comparison between total abstinence and moderate drinking and not excessive drinking; but generally it so happens that once a person takes to drinking, in due course of time the limits of moderate and excessive drinking are obliterated. Social drinking (for companionship) can, many a time, result in regular drinking habit.

Those who approve of drinking in moderation advance some of the following arguments:

1. Drinking is a symbol of modernity.
2. Among Christians, drinking is an integral part of religious customs and of normal life, especially on festivities.
3. It is argued, what is wrong if a soldier takes a peg or two of whisky while performing his duty on a snow-clad mountain post at a height of say about 15000 feet? Similarly, there is nothing wrong if a washer-man, who has to do his work in India by standing in pond or river-water for hours together even on chilly winter-mornings, to consume a peg or two of a drink.
4. It is believed that an occasional peg of whisky contributes to the efficient working of the heart. Proponents of this view possibly ignore the effect of alcohol on the liver.

It would be pertinent to allude to the reaction of the respondents belonging to the medical profession on the statement (G-3) pertaining to drinking.

Response-Pattern

Option	*No of persons agreeing*	*No of persons disagreeing*	*No of persons who did not express any opinion or were undecided.*
(a)	8 (40)	–	1
OR			
(b)	10+1 (50+5)	–	5

Thus 55 per cent of the medical practitioners were ready to tolerate drinking in moderation.

Figures of registered drug addicts, as revealed by the Secretary General of the Association for Social Health in India, New Delhi, indicate that 39 per cent of them started taking drugs for curiosity sake only and 3596 when they were compelled to do so in the company of their friends.

If these figures are any guide in respect of drinking also, there is hardly room for any doubt that moderation in drinking and social drinking can later lead to regular drinking and even to addiction.

The following beer consumption statistics make a useful reading.

World Beer Consumption per Person over the Age of 15 Years

S.No.	*Region of the World*	*Consumption 1960*	*1981*	*Increase per cent*
1	*2*	*3*	*4*	*5*
1.	Central and Western African countries	8.2 ltrs.	61.1 ltrs.	52.9 ltrs. Or 64.590
2.	North America	84.5 ltrs.	120.7 ltrs.	36.2 ltrs. Or 43
3.	Countries of South America	20.5 ltrs.	46.4 ltrs.	25.9 ltrs. Or 12690
4.	Countries of Central America and the Caribbean	12.1 ltrs.	35.7 ltrs	23.6 ltrs Or 195
5.	Countries of East Asia.	3.7 ltrs	24.4 ltrs	20.7 Or 560

These statistics lead to the following conclusions:

1. The industrialised countries outdrink the developing countries.
2. The percentage increase in the consumption of beer in the underdeveloped and developing countries is however, much more than in the developed countries.

If drinking habit is to be cued or to be kept under control, people must be educated on the harmful effects of drinking. In this regard, light may have to be thrown on some of the following aspects in the educational institutions.

1. Addiction – a drag on economy and a social evil.
2. Falseness of the notion that drinking relieves a person of his frustrations and tensions.
3. Drinking as a factor for inducing sleep or sleeplessness.
4. Effect of drinking on appetite.
5. Effect of drinking on various organs of the body, including nervous system, brain, human hearing system, eyesight, sense of smell, etc.
6. Diseases caused as a result of drinking.

Besides, facilities for the availability of alcoholic drinks may be reduced by decreasing the number of working days of the stores selling drinks, reducing their working hours, cutting down the production of alcoholic beverages, etc. Alcoholic drinks may have to be made costlier. But great-foresight and caution will have to be exercised while introducing these or other similar measures so that in the process of imposing curbs illicit practices are not encouraged.

The statement on this subject is below:

"Mahesh and Rakesh were exchanging views on smoking. Mahesh said that one should never smoke. Rakesh said that limited smoking for the sake of keeping company should not be looked down upon. Who in your opinion is right ?

1. Mahesh, who totally opposes smoking

   Or

2. Rakesh, who favours smoking within limits for the sake of company." (S-2)

The opinion-chart on this statement indicated as follows:

Non-smoking Vs Smoking within limits

Option	*No. of persons agreeing*	*No. of persons disagreeing*	*No. of persons who expressed no opinion or were undecided*
(a)	200 (74.9)	5 (2.2)	6
OR			
(b)	54 (20.2)	2 (0.8)	(2.2)

Thus there was a high score in support of the statement suggesting total opposition to smoking.

The composition of tobacco smoke is complex (about 500 compounds have been identified in it) and varies with the type of tobacco and the way it is smoked. The chief pharmacologically active ingredients are nicotine (acute effects) and tars (chronic effects).

Smoke of cigars and pipes is alkaline (p^H 8.5) and nicotine is relatively unionised and lipid soluble so that it is readily absorbed in the mouth. Cigar and pipe smokers thus obtain nicotine without inhaling (they also have a lower death rate from lung cancer).

Smoke of cigarettes is acidic (pH 5.3) and nicotine is relatively ionised and insoluble in lipids. Desired amounts of nicotine are only absorbed if it is taken into the lungs, where the enormous surface area for absorption compensates for the relative lipid insolubility. Cigarette smokers therefore inhale (they have a high rate of death from lung cancer). The amount of nicotine absorbed from tobacco smoke varies from 90 per cent in those who inhale to 10 per cent in those who do not.

Tobacco smoke contains 1-5 per cent carbon monoxide and habitual smokers have 3-7 per cent (heavy smokers as much as 15 per cent) of their haemoglobin as carboxyhaemoglobin, which cannot carry oxygen. This is sufficient to reduce exercise capacity in patients with angina pectoris.

One of the harmful constituents of tobacco is nicotine. This element has been named after the French diplomat J. Nicot, who introduced tobacco into France in 1860. Nicotine is the poisonous alkaloid extracted from tobacco as oily liquid. According to one observation, "If nicotine of two cigarettes in pure form is injected into the system of a human-being, he would die in a short while, as he would when hit by a gun bullet."

Some people regard tobacco more harmful than liquor, one of the important reasons for this being that a smoker goes on consuming nicotine during most part of his waking hours. Other harmful effects of tobacco which students need to be apprised of are as follows. Tobacco is harmful for every part of the body. It reduces one's age. "The time by which a habitual smokers' life is shortened is about 5 min per cigarette smoked." Dolland and his colleagues in 1951 sent a questionnaire on smoking habits to all British doctors. Their respondents' evaluation, after 20 years of 34440 smokers, revealed as follows:

1. Of these 10072 had died.
2. Death rate of cigarette smokers was twice that of life-long non-smokers.
3. Smoking caused death chiefly by:
    a. heart-disease,
    b. lungs cancer,
    c. chronic obstructive lung disease;
    d. various vascular diseases.

Lung cancer accounted for 37,000 deaths in UK in 1974. Medical opinion indicates that if coal-tar content in cigarettes is reduced, deaths from lung cancer would fall by 80 per cent within 20 years.

The risk of smokers developing mouth and throat cancers is 5-10 times greater than that for non-smokers. It is as great for pipe and cigar smokers as for cigarette smokers. Cancers of pancreas, kidney and urinary tract are also more common in smokers.

A smoker has ten fold chances of contracting a heart disease as compared to a non-smoker. If nicotine is consumed by ladies who use contraceptives also, their chance of heart-disease multiplies further. According to one opinion, tobacco smoke contains 2000 harmful elements. Smoking leads to cancer of the larynx, mouth, and oesophagus as well as to lung-cancer. Bronchitis, emphysema, ulcers as well as circulatory diseases are related to smoking, not to speak of many fires which are caused due to smokers' carelessness. By consuming one packet of cigarettes every day, the smoker accumulates 8411 cubic centimetres of coal tar in the system by the end of a year.

Among the causes leading to coronary heart disease (CHD), the most important were cholesterol in the blood, high blood pressure and cigarette smoking. Under the age of 65, smokers are about twice as likely to die of CHD as are non-smokers and heavy smokers about 3½ times as likely.

The obstructive syndrome is as specifically related to smoking as is lung cancer and poses relatively more risks to life than even lung cancer.

In view of the above, laws make it incumbent on the manufacturers of cigarettes to display a warning on the cigarette packs that the product is injurious to health. Canada has perhaps the most stringent laws in the world on tobacco advertisements. Certain contraventions of this law carry a penalty upto 300,000 Canadian dollars.

There are various forms in which tobacco may be consumed by a person.

1. *Sniffing:* Powdered, fermented and scented preparation of tobacco is taken by inhaling. This may contribute to nasal cancer.
2. *Chewing Tobacco:* It is more convenient than smoking a pipe, cigar or a cigarette as no lighting is required in this case. It exposes one to the risk of mouth cancer.
3. *Cigarette Smoking:* Tobacco is rolled inside a piece of paper, which is pasted and thereafter cut into requisite sizes. The spread of the practice of sniffing tobacco in the west is attributed to the 18th century, that of tobacco chewing to the 19th century, of cigarette-smoking to the 20th century, particularly to the 1920s – the period following the World War I. According to the *Encyclopaedia Britannica,*

"In the 20th century, statistics showed that the practice of smoking continued to increase, particularly among women and young people".

4. *Pipe Smoking:* According to the *Encyclopaedia Britannica,* "The smoking of tobacco through pipe is indigenous to the Americas and derives from the religious ceremonies of ancient priests in Mexico. Farther north, American Indians developed ceremonial pipes – the chief of these being calumet or pipe for peace. Such pipes had marble or red steatite (for pipestone) bowls and ash stems about 30 to 40 inches long and were decorated with hair and feathers. The practice of pipe smoking reached Europe through sailors who had encountered it in the New World."

In the Indian rural situation, at times, a young one is initiated into smoking when he is repeatedly asked to prepare *'hukka'* for his parents and grandparents and inhale it too in order to ensure that the tobacco in contact with ambers is ready for inhaling.

According to Laurance and Bennet, one-third of regular smokers began smoking before they were nine. The earlier in life a person starts smoking, the greater is the risk to his health and the life span. If parents smoke, their children are liable to follow their example. Smoking is found to be less common in schools where teachers are non-smokers. If a person reaches 20 without smoking, he is less likely to take to this habit thereafter.

Sigmund Freud was a life-long tobacco addict. He suggested that some children may be victims of a constitutional intensification of the erotogenic significance of the labial region, which, if it persists, will provide a powerful motive for smoking.

About two decades ago, Dr. James Mold, an employee of a cigarette manufacturing company in America, discovered a catalyst which he claimed could protect the smokers front contracting cancer. He said, "we are ecstatic over the fact that we had done something that no one in the world had been able to do."

However, the company's legal adviser counselled his clients not to market the product, otherwise they would be sued for selling something which they themselves have now proved was injurious to public health and they may have to pay compensation for it. So the company shelved the proposal. "God alone knows, how many cancer cases (were) promoted and 'murders committed' by not implementing Dr. Mold's discovery," writes Khuswant Singh.

1. Educating the public through educational institutions, mass-media, voluntary agencies and particularly through associations of medical experts on the harmful effects of smoking will be an effective measure in this regard. The media can play a useful role in weaning the people away from smoking.
2. Students may be apprised of the damage that nicotine, consumed in various forms, can cause to the human-system. Tranquillisers, drugs, acupuncture, nicotine chewing-gums, etc. are not very effective aids to wean people from smoking. A strong will to stop smoking and long course of education on the harmful effects of smoking are the best aids to decrease the incidence of smoking. They should be told that there is no virtue in smoking for company's sake as there is none in committing suicide for company's sake.
3. Smoking should be banned in all closed public places such as offices, buses, underground railway stations, cinema halls, restaurants, etc., as has been done in the USSR. In some cities of India, smoking is banned in buses owned by the government; but this ban is observed more in breach than in observance. It would be appropriate to mention here an Australian episode; "A bus driver, Carroll, 59, an employee of the Melbourne Transit Authority, sued the authority for a compensation as he had contracted cancer due to his exposure to passive smoking on account of smoking by passengers in the bus, though smoking was banned in public transport in 1976. The court awarded a compensation of 52,000 Australian dollars to the victim.
4. The far-reaching effects of passive smoking should also be publicised among the people. Not many people are aware of the fact that they are liable to suffer from the harmful

effects resulting from active smoking if they keep close company with the smokers, notwithstanding that they may be non-smokers.

5. The students and the general public should be told about the extent of paper which cigarette wrappings consume. This, in turn leads to depletion of forests and importing of more paper from abroad against foreign exchange.
6. The cigarette manufacturers should be banned, by law, to use the name of gods or eminent persons as their trade mark such as Ganesh Bidi.

Role of Sex: Sex is the basis of all life. The different mythological theories of creation write about various sexual acts such as of the sky and the earth. It is held that Eros (i.e., sexual love) or 'Kama' was one of the first creations of the universal egg from which all creation followed, though Adam is believed to have been expelled from Heaven on account of not having been able to control his sex instinct. Kama (sexual instinct) has been described in the ancient Vedas and the subsequent works as all pervading and invincible. It is present in human beings, animals, birds, creepers, trees, fruits, flowers and, in short, in all life forms. During the period of youth, its impulse tends to become irresistible.

Matrimony is essential for the evolution of life. Numerous sages and great personalities tasted the joys of married life. If they did not, how would it be possible to propagate and perpetuate the genes of good people in this world! In Indian history, there is a mention of so many rishis having led a married life before becoming ascetics. All this shows that sex instinct and the urge to multiply is natural and universal in all humans.

Pre-Marital Sex: Post-marital sex being a normal phenomenon of life, a statement to study the attitude of the society only to pre-marital sex was included in the questionnaire. As pre-marital sex indulgence concerns the unmarried persons only, this statement was included in the questionnaire for students. This statement was as follows.

"No stigma should be attached to sexual relations before marriage". (S-9)

The reaction of the respondents is given below:

Pre-marital Sex Indulgence

S.No 1	Category 2	No. agreeing 3	No. undecided 4	No. disagreeing 5
(1)	Total population (267)	58 (21.7)	27 (10.1)	182 (68.2)
(2)	School students (M)	14	14	42
(3)	School students (F) (86)	8 (9.3)	9 (10.5)	69 (80.2)
(4)	College students (M) (34)	20 (34)	1 (2.9)	13 (38.2)
(5)	College students (F) (70)	12 (17.1)	3 (4.3)	55 (78.6)
(6)	African students (7)	4 (57.1)	– (42.9)	3 –

The preceding table indicates that college students approve of the idea of pre-marital sex to a higher degree than their counterparts in schools. This maybe attributed, to an extent, to the demand of the age. Girls seem to be opposing the suggestion of pre-marital sex much more than the boys, including those who are in college. One of the reasons for this could be that the girls realise that the stakes involved for them in pre-marital sex relationship are far greater than those for boys. Pre-marital sex relationship especially if it results in pregnancy, can ruin the subsequent married and family life of the girl or even the chance of marriage itself. It is interesting to note that in a tradition-bound society, which taboos pre-marital sex, a large proportion of youths seem to be favouring it.

Pre-marital sex relationship may be of three kinds:

1. When both partners are unmarried.
2. When one of the partners is married.
3. Homosexual indulgence.

Views differ widely on this statement.

According to Waltermarck, "Among many uncivilised peoples, both sexes enjoy perfect freedom previous to marriage, and in some cases it is considered almost dishonourable for a girl to have no lover.

The East African Bares and Kunama do not regard it as in the least disreputable for a girl to become pregnant nor do they punish nor censure the seducer. (Among the Wanyoro) it constantly happens that young girls spend the night with their lovers, only returning to their father's house in the morning, and this is not considered scandalous. (In Solomon Islands), female chastity is a virtue that would sound strange in the ear of the native. Among the Angami negros, men are desirous before marriage to have a proof that their wives will not be barren – Chastity begins with marriage.

"In some cultures, (pre-marital sexual) intercourse is believed to be necessary if pre-adolescents are to mature sexually. The Trukese of Caroline Islands build small huts especially for this purpose. Among the Alorese of Indonesia, mothers routinely masturbate their children in order to pacify them."

About half a century ago, Kinsey studied sex before marriage in America. His interviews between 1940-48, with some volunteers who agreed to cooperate with him in his study, included members of the social club as well as persons in prison (who of course can not be regarded as a true sample of the general society), revealed 85 per cent of all Americans had experienced pre-marital sex, 70 per cent had visited a prostitute and over one-third had participated in at least one homosexual act. Kinsey's data suggest that a wave of sexual liberation occurred much earlier than is usually believed, probably in the generation that came of age after World War-I . His study found that only 8 per cent of white women born before 1900 had pre-marital intercourse before the age of 20, but that among those born between 1910 and 1929 the figure was 22 per cent. Later surveys indicate that the second wave of liberation occurred in the second half of the, 1960s and the early 1970s. Kinsey found that two-thirds of the single women aged 18 to 25 in his sample were virgins, while Hunt's survey, conducted in the 1970s found that only one-fourth of the single women in this age bracket were virgins.

Among Muslims: Shiite school in Islam believe in Mut'ah, which is defined as a fixed-term marriage before the regular matrimonial alliance. Its main features are as follows:

1. A man and a woman may enter into a matrimonial contract for a fixed period, which may or may not be renewed. It may lead to a regular marriage as well.
2. In this temporary kind of marriage, man has no obligation to pay the expenses of the woman as in the case of a regular marriage.
3. In a regular marriage, woman is to accept man as the head of the family, but in the case of Mut'ah, she may not accept him as the head of the household.
4. Any of the two partners can exercise birth-control without the consent of the other. In Islam, birth control is permitted before conception takes place; but abortion is not allowed in Islam.

This custom of legalised pre-marital sex has both pros and cons.

Pros: By the time a girl or a boy reaches 18 years of age, sex instinct develops in the individual; but social maturity and financial independence is yet to come about. Therefore, regular marriage cannot be recommended in such circumstances. On the contrary, celibacy can also create psychological and emotional problems. Therefore, to avert chances of 'sexual communism', in which case any boy can mate with any girl, Mut'ah is recommended which puts some limits on the pre-marital relations.

Cons: The opponents of Mut'ah argue that this system amounts to licensing and codifying sex-experience prior to marriage. Further, sex indulgence is a chain-reaction, and depending on the opportunities available, young boys and girls can become victims of promiscuity. Whereas Mut'ah may satisfy animal instincts in a man or a woman, it militates against mutual loyalty.

Chastity: Contrary to this, chastity is prized as a great-virtue by many societies. Hindu ethics as well as that of the Christians lay emphasis on self-mortification and asterism, whereas Islam refuses to accept that sex-abstinence is a necessary constituent of high morals. Speaking before a rally of 400,000 young people in Santiago DE Compostela on August 20, 1989, Pope JOHN Paul warned the youths against the dangers of easy sex, saying that a materialist society had robbed love of its real content. He added that modern ways of life had reduced "love to a temporary experience of personal gratification and even of mere sexual enjoyment. How can you not recognise in all this the worm of a consumist mentality which has slowly emptied love of its transcendent content?" (The Statesman, New Delhi, 21-8-89, p.8)

Among the Moroura tribe in Australia and in Nias, sexual intercourse before marriage is punishable with death. Even among the Muslims, according to the religious laws rape before or after marriage is punishable with death.

Among several civilised nations, different standards of sex morality are applied to men and women. Chastity is considered a male virtue and a female duty, which obviously, amounts to having double standards in sex matters.

In the balance, one maybe inclined to think that the temporary marriage recommended in Islam would be better than the irresponsible sex-indulgence or what is sometimes termed as communism in sex and also better than inhabitation with prostitutes or members of the same sex; but this is not all. It has several negative factors also. Man always desires to marry a virgin. Even the college students who, expressed no objection to sex before marriage would not be inclined to marry a non-virgin girl. The offsprings of unmarried mothers are a social problem. Also, it hurts the life of a child who is born as a result of pre-marriage intercourse if in the pre-marital sex experience one of the partners is married, it becomes a case of conjugal infidelity. In this case also, if the woman becomes pregnant, the problem of illegitimate child becomes equally grave.

Teenage Pregnancy: When we talk of pre-marital sex, we should be conscious of the gravity of teenage pregnancy. An article appearing in the World Health Journal (April-June, 1988) on Teenage Pregnancy in the America by CESAR A. Chelala (pages 22-23) is revealing. The writer calls it a new epidemic of modern times. "In Latin America as a whole, more than 20 per cent of pregnancies occur in women below 18 years of age. In the Caribbean, almost 50 per cent of all teenage births occur to women of 17 years old or younger. In the United States, close to one million adolescents between 15-19 years and 30,000 girls below 15 become pregnant every year. True estimates of pregnancy are lacking because at least 20 per cent of pregnancies end in abortion, which is illegal throughout the region (except Cuba and USA). The early and great frequency of sexual activity among teenagers seems to be a critical factor. Many girls become pregnant because they are ignorant about the basics of the sex act or the birth control. "............a US national survey estimated that almost 70 per cent of pre-marital pregnancies among teenagers were unwanted". " One US study among 180 pregnant students found a prevalence of attempted suicides of 13 per cent".

Pregnancies are likely to be more in the case of girls in lower socio-economic groups where both parents are working and do not have enough time to spend with their children.

Adolescent pregnancy has physical and psychological consequences – lack of proper medical care, high child / mother mortality rate, more still-births, clandestine abortions risking lives, education getting hampered.

"The teen age pregnancy rate in India varies between 8 to 14 per cent. The Wardha study of 1000 pregnant teenagers, who came for abortion, noted that 88 per cent of the girls had no knowledge of the possible consequences of sexual intercourse."

With the increase in the population of youth and the family-planning programmes being regularly advertised all over, the curiosity of the youth to get acquainted with sex matters gets accentuated. If the youth are not given, the information which they desire to seek in this regard, they would try to secure information on this subject through channels which may be unreliable and misleading. It therefore, appears essential to disseminate an authentic information on sex matters to the youth through informed channels.

Sex Education: In regard to sex education through schools and colleges, there was one statement in the questionnaire for students and one in the questionnaire for general non-student population. These statements are reproduced below:

"Sex education should be imparted in:

1. Schools;
2. Colleges." (S-12)

"Introduction of sex education would be desirable:

1. In colleges as well as schools

   OR

2. In colleges only

   OR

3. Neither in schools, nor in colleges" (G-12)

Students' response to the above statements is indicated in the following charts.

Sex Education (S-12)

S. No.	Category	Statement	Number Agreeing	Number Undecided	Number Disagree-ing	Number of persons who expressed no opinion
(1)	Total student population (267)	12(a)	157 (58.8)	49 (18.4)	60 (22.5)	1 (0.3)
(2)	Total student population (267)	12 (b)	166 (62.2)	77 (28.9)	18 (6.7)	6 (0.2)

Sex Education (G-12)

S. No.	Category	Statement	Number Agreeing	Number Undecided	No.Dis-agreeing
1.	Total population in general on-students category (881)	(a)	419 (47.6)		4
		OR			
		(b)	303 (34.3)	21 (2.4)	1
		OR			
		(c)	133 (15.1)		

In the case of the general population, only 15.1 per cent of the respondents felt that sex education should neither be imparted in schools nor in colleges; the rest were in favour of introduction of sex education in the educational system at some stage. As regards students, larger number of them favoured sex education at the college stage. Surprisingly, the college students (M) did not support this idea to the extent the college students (F) did. Further comparison between the extent of agreement expressed by school and college students in case of statement (S-12) of the questionnaire brings out the following picture:

Comparative Study of the Opinions Expressed by School and College Students: (S-12)

S. No.	*Category*	*Statement*	*Extent of Agreement*	*Statement*	*Extent of disagreement*
1	2	3	4	5	6
1.	School students (M) (70)	12(a)	38 (54.3)	12(b)	37 (52.9)
2.	School students (F)	-do-	36(51.8)	-do-	48 (55.8)
3.	College students (M) (34)	-do-	27(79.4)	-do-	23(67.6)
4.	College students (F) (70)	12(a)	49 (70)	12(b)	53 (75.8)
5.	African College students (M)	-do-	7(100)	-do-	5(71.4)

While imparting sex education, one has to have due regard to the age-level of the students and their total environments. The students in cities,, possibly, have a greater need for sex education than their counterparts in villages whose exposure to the exhibition of sexy situations is less. This happens because the invasion of western culture, sex-stimulating films and advertisements is more on the inhabitants of the cities than those living in villages.

Role of Mass-Media: In any educational programme, mass media have an important role to play. Therefore, while considering the type of educational programmes on sex-education, it may have to be examined how far some of tile pictures exhibited on the silver screen and the material appearing in the print media would help or hinder the desirable sex-education process. The statement pertaining to this aspect, as given in the questionnaire for the general population, was as follows:

"There is not much harm in exhibiting pictures with sex-appeal in newspapers, on TV and cinema screen" (G-4)

Out of the 881 respondents, only 16.6 per cent agreed with this statement, whereas 695 of them (i.e., 78.9 per cent) disagreed and 4.5 per cent of them were undecided. Though the wording of this statement had the following flaws, yet the high percentage of 78.9 per cent makes the verdict of the respondents very dear. These flaws are:

1. the use of 'not much' make the statement somewhat elusive;
2. the term 'sex-appeal' is used in good sense also, though in this case its import and the Hindi translation of the statement made it clear that' sex appeal' in this context meant 'sex-stimulating situation'.

Curriculum: The curriculum for sex-education could be drawn up on the following lines:

(a) Pre-school Stage – Home Education

1. At the pre-school early stages, the child desires to know how he came into this world. He also derives some pleasure in play with his genitalia.

It is for the parents and not the school to tell the child that he came out of the mother's tummy. Mother can also show some animals or women who are pregnant to make her explanations simpler. As regards playing with genitals, the education has to be practical, i.e.., by covering the child's organs with napkins or panties so that the child does not develop this practice.

(b) School and College Curricula

Education at the school and subsequent stage may include the following information:

2. Development of egg cells in the body among animals which do not lay egg.
3. Emotional ties between young ones and the family in humans as well as animals.
4. Secretion of sex hormones and their effect, menstruation cycles in girls, wearing of sanitary pads, pain during menstruation and other allied problems and their precaution, pregnancy, semen formation in boys and wet dreams.

5. Causes of artificial sex stimulation -such as obscene literature, excess consumption of tea, tobacco, coffee and spicy foods.
6. Masturbation and its harmful effects.
7. (Near adulthood) Coitus act resulting in pregnancy, danger to life, illegitimate children, abortion, contraceptions, promiscuity, prostitution, veneral diseases.
8. Homosexuality and its harmful effects. Harmful results of pre-marital sex, need for sex-abstinence, effect of extra-martial sex on the harmony of family life.
9. Causes and Consequences of AIDS

Sex education can be imparted more easily through subjects like physiology, biology, nature-study, etc. Sex education should also include the question of sex ethics and advice on a balanced, responsible and happy family life. With the weakening of the family ties after the break-up of the joint family system, particularly in the urban area, the need for 'family education' has tremendously increased. Some of the Indian temple architecture, like that of the Konark Temple in Orissa, and stone sculptures and paintings elsewhere can provide a very good material for education on this erotic subject in a sublime manner.

Some people feel that sex is instinctive and no education is necessary in matters of sex. Such persons, perhaps, take a limited perspective of sex-education. There is much more in sex than sex alone- the aesthetics, the dedication, and so on. Therefore, instruction in sex matters is essential. As any discussion on sex has been treated in the Indian society, for centuries now, as an area of secrecy, the teachers charged with the responsibility of imparting sex education will have to be given a thorough training in this subject, both in content and pedagogy, so that they can debate various issues pertaining to sex in a normal way and with natural ease and without getting embarrassed in any manner.

Political Features

The school is the miniature society and a social institution to serve its purposes. It is entrusted with the responsibility of training and bringing up the students so that they may be able to take part effectively, efficiently and harmoniously in the community to which they belong. Prof. K.G. Saiyidain has rightly remarked, "A people's school must obviously be based on the people's needs and problems. Its curriculum should be an epitome of their life. Its method of work must be approximate to theirs. It should reflect all that is significant and characteristic in the life of the community in his natural setting". Hence the schools as an epitome of people's life, must include all kinds of useful activities, important problems and significant features of our day-to-day life.

Sociological Influences: Students should be given various learning experiences inside the school which they will face outside. Otherwise, they will prove worthless, helpless and ineffective individuals. On the other hand, if they are properly trained in the schools, they will be able to occupy their rightful places in the community. The need for linking the school life with the life in the community cannot be overemphasised. The Secondary Education Commission, 1952 have stressed, "The starting point of educational reform must be the rethinking of the school to life and restoring of the intimate relationship between them which has broken down with the development of the formal tradition of education". Such intimate and living relationship between the school and the community can be established by introducing various kinds of active occupation, the desired values and behaviour patterns in the school.

The entire school life should be infused with the spirit of community life. Dewey held that the school be a child's habitat where he should learn "through direct living, instead of being only a place to learn lessons having an abstract and remote reference of some possible living to be done in future. It gets a chance to be a miniature community, an embryonic society". The relationship between the school and the community can be summed up in the following words of Reburn, "There must be vital connection between the life of pupils in schools and the life of the community from which they come. There must be a vital connection between the school, which is the corporate life of pupils and teachers and the community.

Relationship between the school and the community can be strengthened and the former can be made a miniature community through close corporation and mutual give and take between them. In a democratic set-up such relationship is not only essential, but also imperative. The community, even at the stage of its underdevelopment, possesses abundant resources like forms and forts, old and new constructions, temples and monuments, places of new social, economic, historical, cultural, technological, artistic and aesthetic importance. Besides, there are fairs and festivals, rich heritage and customs as well as rich human resources like artists and artisans, craftsmen and clergymen, doctors, teachers and so on. On the other hand, schools have buildings, equipment, furniture, playground, libraries and laboratories. They are not merely meant for children. The whole community must benefit from these resources.

Educational literature frequently makes reference to "family back ground" as a summarising variable that impacts on school achievement, level of aspiration, and other learners outcomes associated with the school experience. This practice has become so common as to create stereotyped expectations associated with family background, social class identity, and family structure. For example, when a teacher speaks of a child coming from a "good home," this is usually interpreted to mean that the child is the product of a two-parent, middle-class, white family of adequate economic means who are loving and supportive of the child. By implication, variations in this stereotype mean that the home is less "good." This generalisation is, of course, often incorrect. A child can be the product of a middle-class family and have a psychologically and emotionally devastating family experience.

A child from a working-class family might have a family background that provides strong psychological and emotional support. A child from a one-parent family can be and often is secure, stable, well-adjusted, and a high an achiever. The number of children coming from disadvantageous families is so high today that it can no longer be considered in any sense unusual. The teaching profession would be well served by dropping the use of the expression "broken home" because of its pejorative and inaccurate connotations. The human family is a small kinship system, which means that it helps us identify who is related to whom by blood or by marriage. The smallest family unit is the conjugal family and includes only the husband and wife. If children are added to this basic unit, it is called a nuclear family. If, within the same household, there reside other relatives besides the father, mother, and children, we have an extended family. Extended families usually include parents, children, grandparents, aunts, uncle, cousins, and possibly great-grandparents.

In modern, urban-oriented societies, and especially in the United States, the nuclear family is the dominant arrangement. In the traditional societies of the world, some form of the extended family is the more common pattern. The family is what sociologists refer to as a "primary group," and it is the most significant instrument of socialisation in lives of most persons. It is a "primary group" because there is great deal of face-to-face, direct contact among members and because much of the individual's life is spent in such a group. It is from the family that the individual learns some of the most important things he or she needs to know in life.

The family teaches the individual the traditions, folkways, and mores of society to speak a language, to develop a value system, and to form conscience. How the individual views himself or herself and the world is largely conditioned by family life. There is convincing research support for the idea that social class influence of the family is felt early in life. Kagan reports several studies conducted in recent years that document the early – as early as the first birthday – and persistent influence of family social class identity. He reports studies indicating that "by 20 and 29 months of age, the class of the child's family had become a major predictor of attentiveness, volcanisation, and smiling." One of the most emotionally relevant differences associated with family social class identity is in the area of language development and use. These differences have been consistently confirmed by studies that extend over a period of forty or more years.

The differences on whatever measures are taken uniformly favour the middle social class children over those from lower or working-class families. The middle-class child typically has learned both restricted and elaborated linguistic modes and is able to adjust to the appropriate mode as the social setting varies. The lower-class child has his or her communication limited to the restricted linguistic mode. Middle-class children have more extensive vocabularies, produce more mature sentence structures, and are more precise in their articulation and pronunciation than are lower-class children. These differences seem to be associated with early family upbringing in which mothers encourage cognitive development by talking to their young children, providing explanations, encouraging freedom of exploration, and avoiding unnecessary intrusions into the activities of the child.

The middle-class child's socialisation provides more freedom, more self-involvement, more of an attitude of being able to control one's destiny. The environments of lower-class families are less conducive to this type of cognitive development, especially in language development. These differences are related directly to the child's performance in school. The sequence of achievement disability proceeds somewhat as follows: The child comes to school with limited linguistic facility and with restricted first-hand experience out of which concepts can be intellectualised. These deficiencies become immediately apparent when the child is confronted with the task of learning to read, usually in the first grade.

The school becomes largely a custodial centre rather than an educational institution. Teachers have an incredibly difficult, stressful, and tension-ridden work environment. For many it is a challenge simply to survive in what is literally a dangerous place. Student absenteeism is high, and school officials make little effort to enforce truancy laws. Teachers usually breathe a sigh of relief when the troublemakers are absent. The school curriculum is based on values that have little meaning for children and teenagers who come from a world of strutting pimps, drug pushers, and apartment houses shared with rats and cockroaches. A child terrorised by a street gang can hardly be expected to be highly motivated to attend to what is being presented in the classroom. Teachers, school officials, political leaders, legislators, and citizens generally must realise once and for all that the problems associated with the underclass cannot be solved through educational efforts alone.

To suggest that public education can somehow provide solutions to a dangerous social condition of this magnitude is hopelessly naive. Indeed, public education by itself is no answer at all to this serious cluster of problems. If this society is to overcome the problems of the underclass, it will have to mount a major rehabilitation effort, of which education would necessarily be a part. But national priorities are not so arranged at the present time, and there is little to suggest that they will be in the near future.

Meanwhile, the most prosperous and highly educated nation on earth continues to sit on a social time bomb that will one day surely explode and create a major upheaval in social relations in this country. What is being suggested should not be construed to mean that the school has no constructive role to play in helping children from the underclass. We are suggesting only that the school cannot do it along. Successful strategies for dealing with the problems of the underclass will need to extend across a broad spectrum of social and health services, including social, psychological, economic, political, educational, vocational, and allied health spheres.

Until society sees to fit to tackle this problem on a broad front, teachers must do the best they can in making a conscientious effort to teach these children to read, write, and perform basic arithmetic operations simply as survival skills. Beyond that teachers need to identify and help develop whatever talents these individual children have. Additionally, the school can contribute to finding solutions to the problems of the underclass by better educating the children of the middle and upper classes, who will one day be the policy-makers who set social priorities for the society.

Political Influences: Today, school governance was thoroughly enmeshed in political activity at the local level. Corruption, graft, and less objectionable forms of political influence characterised hiring of teachers, purchase of textbooks, construction contracts, and the employment of non-certified personnel not only in the large urban areas but in smaller communities as well. The malaise that afflicted school administration was so pervasive that it became the object of a national reform movement, consistent with the progressive political ideology of the early part of this century. Highly respected and powerful scholars and educational leaders gave support and leadership and the effort to remove school governance from the girl of local politics.

As a result, school districts came to be autonomous political units, independent of other local jurisdictions such as local country or city governments. The efforts to reform the administrative structure of schools also initiated the movement to consolidate and centralise schools, a process that continued well into the 1970s. The consolidation and centralisation of schools appealed to business-minded Americans because it resembled a cororation structure. The efficient corporate model was obviously attractive as compared to the decentralised unbusinesslike way in which schools conducted the affairs. Indeed, the term 'businesslike' is considered to be synonymous with efficiency. A major cornerstone of the school district reorganisation effort was eliminated wherever possible those districts that were too small to operate efficiently.

At the end of World War II there were over 100,000 school districts in the United States. By 1960 that number had been reduced to 42,000; and by 1970, to about 18,000. Today there are approximately 16000 school districts. The legal governing body of school district is the school board, known as the board of directors or as the board trustees.

School boards are given the power by state law, established local school policies, consistent with guidelines and laws of the state. Their responsibility, include such matters as hiring the superintended approving curriculum documents, setting the school calendar, negotiating contracts with employees, receiving and expanding funds, establishing a budget and developing and maintaining facilities. The school superintendent is the chief executive officer of the board, and it is through the office of the superintendent that the board policies are actually implemented.

School board members have authority only when sitting as a board; that is, an individual school board member, on his or her own, has no more authority to direct the affairs of the school than any other citizen in the community has. This is often not understood; and, as a consequence, sometimes board members will attempt to usurp responsibilities that are more appropriately those of the superintendent, principal, teachers, budget director, curriculum director, or other professional staff. Many studies have been made of the nature and composition of school boards. Typically, boards consist of five to nine members who are elected for a period of three to six years. Their level of education tends to be above average, which means that school policies in most places are being set by persons who have themselves succeeded in school.

Most board members are proprietors, businessmen, executives, and professionals. Tradition and public policy hold that school board members should give their time and talent as a public service; that is, they should serve without salary or only at a token salary, they should be non-partisan in their political affiliation, and they should be elected at large. These requirements are intended to reduce the more objectionable political aspects of board membership.

In recent years there have been renewed efforts to halt or even to reverse the trend towards centralisation. This has been promoted by minority groups, especially blacks, living in the large urban areas. It would be a mistake, however, to associate this movement only with inner-city blacks. Other groups such as Indians and/ or Native Americans, Latinos, and Asians have also been active in working towards greater local control of schools. The disadvantaged minorities have not been satisfied with the way schools have served their children and fed that, if they had more to say about setting policies and making decisions, they would be better served.

As a consequence, there has been considerable pressure to get what is called "community control" schools. The issues that are involved in centralisation decentralisation sift represent value conflicts having to do with keeping decision making close to the grass root while at the same time having an efficient administrative operation. Because it is essentially question of values, there is no way that it can be resolved entirely on the basis of expertise. What is being sought is a system that will maximise both democratic control over decision making and also that will maximise efficiency in governmental operations.

Most advocates of centralisation would not go so far as to push for a nationally centralised system of education. Summarily, most advocate a decentralisation would not want to return to the days of a school district at every crossroads in the land. Most feel that schools should accommodate the educational needs of local constituents. Two ways have been suggested to achieve school decentralisation. One might be called political decentralisation. Under this system, large consolidated school districts now operating under the direction of a single school board would be split into smaller districts. This, essentially, would reverse the consolidation trend that has been taking place for the better part of this century.

Each of these administrative units would be autonomous, have its own school board, and be under local community control. As such, it would set its own policies, hire personnel, approve the curriculum, set its own budget, and so on. Although this type of political decentralisation has a certain nostalgic attraction, it is really not a practical or feasible alternative to what is done today. Public services for today's communities require such complex and costly delivery systems that a metropolitan-wide effort is required to have them become a reality. This is true in the case of public transportation, water and electrical services, fire and police protection, sewer and garbage services, and, to some extent, health care. Increasingly, such service districts include all, or a good part, of the entire metropolitan region.

The move towards political decentralisation of school districts would take communities in the opposite direction of metropolitanism, a very unlikely development. A more realistic approach to preserving or restoring some semblance of local control is to handle it through administrative decentralisation. Under this arrangement, the large district boundaries are maintained, and there is a central governing board. An effort is made to separate functions in a way that will maximise both efficiency and democratic control. Thus, such activities as purchasing, budgeting, and personnel services are handled at the central administrative office level. This provides the district many advantages in terms of cost and efficiency.

For example, payrolls and book-keeping can be computerised for the entire district, thereby eliminating the need to duplicate the operation several times. The district can exercise economies by making purchases on bid and making them in such large quantities that suppliers are willing and able to offer them at a lower price. Whereas the district centralises functions such as those just described, it decentralises functions that involve curriculum and instruction. For instance, the district might be divided into several areas, each of which is headed by a regional administrator. Such areas might even have school councils that consist of local parents who advise the school authorities with respect to school policies. Although such school councils may not have any legal basis, they do, nonetheless, have credibility and can have considerable power. Parent and student involvement at the local level helps to ensure a meaningful curriculum for children attending local schools.

Because administrative decentralisation preserves the advantages of both a centralised and decentralised system, it has been implemented in many large city school districts in recent years. In a decentralised administrative organisation the authority and responsibility for education is dispersed. For example, in our system, authority and responsibility for education is shared by local communities, larger school districts, and states. The assumption is that when there is such dispersion of authority and responsibility, the system will more adequately take care of local needs. A further assumption is that this system will reduce the amount of arbitrariness in decision making because the farther away from the effects of a decision that the decision is made, the more likely it is to be arbitrary. Another assumption is that local involvement in decision making will result in greater interest in schools.

All of these assumptions and others like them that surrounded the complicated issue of local control are open to much question. They are partially true: perhaps in some instances, altogether true. In other cases, they may not be true at all. Certainly, local control of schools does not necessarily mean better quality of education.

Individuals usually favour local control of schools for some combination of the following reasons:

1. Advocates of local control feel that the schools belong to the people: This view argues that schools are literally owned and operated by the taxpayers of the local community. School administrators, teachers, and other personnel associated with the school district are employees of the local community. At

such, these employees should be accountable to the citizens of the local community. Because the schools belong to the people of the local community, it follows that local citizens would inform the school as to the kind of educational programme they want for their children and would supervise the work of the school to ensure that it was being carried out. Teachers who are unable or unwilling to perform in accordance with the mandates of local citizens would be dismissed. Naturally, organised teacher groups believe otherwise, namely, that they do not work well when they are under such close supervision and that are professionals they must be permitted a considerable amount of autonomy in the exercise of professional decision making.

2. Local control reflects local needs: Advocates of local control often state that they do not need bureaucrats in the central office many miles away or officials in the state education office to tell them what kind of education their children need. They insist that the persons who can best decide on local education needs are local citizens. Obviously, this point and the foregoing one are closely related.

3. School management and curriculum will be more closely monitored when control is at the local level. If this administrators and teachers are accountable to local constituents, parents will more likely monitor school activities themselves rather than rely on representatives from the central office to do it for them. Who is the first to know if a teacher is not doing his or her job responsibility and adequately? Local control advocates say that it is the parents of the children with whom the teacher is working. Consequently, they are able to call such a situation to the attention of the school authorities can be made at the local level. Parents are not likely to be very effective if they have to register their concerns at some anonymous bureaucratic office many miles away. It follows, too, that the schools would be more responsive to such local criticism and suggestion. In order to get this type of representatives, local authorities must remain vigilant and alert to what is going on in the schools.

4. *People who are affected by decisions should be involved in making them:* Of course, those who support local control believe this is more likely to happen if decisions are made at the local level. The fact is that in representative governments we are more often than not involved only indirectly in decisions that affect us. Even though participatory democracy has become a popular concept in recent years, it is simply not possible for us to be involved in everything. Local control advocates would say that one is not asking to be involved in everything – only in the education of one's children. The education of one's children is an extension of parental responsibility and cannot and should not be relinquished to a stag agency such as the public schools. Democratic decision making under a representative system is based on the concept of 'consent of the governed,' and in this case, the governed are not willing to give consent.
5. *People will be more interested in schools if they are controlled locally:* If local communities perceive schools as belonging to them, they are more likely to pride themselves on having good schools and will be more interested in supporting what they do. Schools are more than educational institutions; they are symbols of the community. They contribute to community solidarity and unity. The high school football or basketball game is not only an athletic event; it is an expression of community loyalty and esprit de corps. Thus, the school has important socialisation functions for a community beyond fulfilling educational needs that are served when the school is controlled and operated locally. Often communities are very upset about a school closure, not because this will do violence to the education of children locally but because such a closure says something about the community itself.

Individuals usually do not favour local control of schools for some combination of the following reasons:

1. *Localism promotes parochialism, narrowness of vision:* Opponents of local control do not deny many of the arguments made in its favour. Indeed, they would say that it is precisely what is wrong will be reflected in the school when schools are controlled locally. "If desegregation, for example, were left

in 'the hands of local citizens, we would still be at the stage we were in the early 1950s," say opponents of local control. Opponents also say that children may get their early educations at the local neighbourhood school, but they do not usually spend a lifetime in that community. They grow up and move away. Therefore, they must have a familiarity not only with the local area but with the world beyond.

2. School expenditures will work to the advantage of local elites and to the disadvantage of low socio-economic status families: It is further argued that the social class composition of the community has little to do with contradicting this principle. However the local community is constituted with respect to income, occupation, race, ethnic groups, or social class, some group can be identified as being most influential in decision making. That group of elites will see to it that school funds are expended in accordance with its views of what constitutes a good education. The net effect of this is a replication of social class structure at the local community level from one generation to the next. Local control ensures that the 'haves' will continue to get in generous amounts and the 'have-nots' will continue to be shortchanged by the schools. Opponents argue that if it is equity of education that is being sought, one does not get it through the mechanism of local control. They would say that policy formation has to take place beyond the local community-where local elites cannot dominate and control decision-making process.

3. When education is controlled locally, it does not address itself to large societal needs: The world of today's children does not begin and end at the boundary of the local community. People and places interact with each other. A community has many interlocking social, political, economic, and geographic links and bonds with other places. As a nation, we are heavily involved in international relations. "We cannot educate children only in terms of the needs of the local community," the opponents to local community control tell us. Every school in the country has to be concerned with those broad social issues that concern us as a nation. Today's children not only need a community perspective but

also need to see the world from the perspective of the region and the nation.

4. Localism works contrary to the integration of neighbourhoods into the larger community: Too great an emphasis on the local community can promote separatism just at a time when society is stressing integration. If one community is black, another is Chicano, another all white, and so on, the local, operation of schools in those communities accentuates the racial and ethnic variable. Very substantial efforts have been made in recent years to avoid such homogeneity of racial and ethnic composition of schools; indeed, they are required by law to be otherwise constituted.
5. The improvement of education is too complex, too sophisticated, and too large to be handled adequately at the local level: What is really being said here is that what is needed is not localism but metropolitanism. The concern should be spread over a wider rather than a narrower area because social and educational problems are usually pervasive to a large region, not limited to a neighbourhood, according to the opponents of local control. To believe that local persons, who are often poorly educated themselves, have the wisdom and the expertise to make sound decisions about the education of their children approaches incredible heights of naivete. Such procedures might have been adequate in the day of the schoolmaster and his little rural school, but not for children who will spend most of their lives in the twenty-first century.
6. *Funds for special projects are often used poorly:* Guidelines for the funding of federal education projects have required that projects be governed by local executive boards consisting of members of "the community." Modelled after self-improvement programmes in Communist countries, such involvement is supposed to develop skills of participatory democracy and at the same time ensure the project's relevance to the group involved. Of course, it rarely works that way, according to opponents, because the money is squandered on worthless projects or otherwise misappropriated. Persons who have been totally unaccustomed to handling anything

but trivial amount of money in their personal lives, now find themselves in charge of huge amounts of money, a responsibility for which they are totally unequipped either by personal experience or professional expertise.

7. *It is more difficult for professional personnel to engage in negotiated contracts:* It is obviously easier for a union to negotiate with a single school board of a large city than it is to deal with all units all the local level. But supporters of localism would take the opposite view. They would want to bargain with teachers at the local level. They would want to hire, fire, or reward teachers. This issue, therefore, is one that places the teachers unions and local community advocates on opposite sides.
8. *Localism defeats high priority social goals such as integration, cultural pluralism, and metropolitan coordination:* Great strides have been made in the last quarter century in achieving these broad-based social goals. Localism seems to fly in the face of these trends. It is difficult to support school integration and desegregation, for example, and at the same time support the separation that is suggested by local control. The revitalisation of communities cannot occur through efforts of the local community alone. The problems are too complex and their resolution too costly to be handled by such a limited effort. Good social policies are made by persons who are competent to make them. To suggest that people at the local level have such competence by themselves is probably expecting too much.

The issue of local control of schools is not one that can be resolved entirely on the basis of objective data and expertise. It is obvious that the problem is enmeshed in different and conflicting value orientations. This means that opponents as well as advocates are likely to overstate their particular point of view. For instance, let us take the matter of professional autonomy. Should teachers be accountable to, and supervised by, local laypersons? The answer to this question cannot be a clear-cut yes or no. The truth is that teachers are accountable to local citizens and parents, but this does not mean that teachers are or must be under the constant supervision of parents. Teachers do exercise a considerable amount of freedom is carrying out their duties and responsibilities, yet they are not completely autonomous in performing their professional roles. This same type of compromise position characterises most of the specific points of conflict regarding local schools.

Pressure Groups and their Influences: The traditional views in politics are too narrow for our purposes. Instead, we should take a broad view of politics. The definition of Professor Harold Lasswell will suit our purpose: "The study of politics is the study of influence and the influential." Politics in education, therefore, may take many forms because individuals and groups are always trying to introduce the views of others. Earlier we discussed critics of the schools and, in that setting, talked about various pressure groups. Such groups might also be considered political power groups that attempt to, and often do, influence school decision making.

Teachers and even administrators have held the belief that schools somehow are outside the realm of politics. Despite the fact that the most important decisions affecting schools, namely, how they are financed, are wholly political, school people insist on clinging to the naive belief that "schools are above politics." As a result, persons most closely associated with school decision making often have the least well-developed political skills. The time has come for those associated with school decision making to set aside the romantic notion that schools are not involved in politics. Schools are and always have been influenced by local, state and federal political power groups. This is so because schools serve many constituent groups, each of which has its own set or priorities for the schools. They are not all mutually exclusive, of course, but neither are they wholly alike. The efforts to influence decisions in accordance with a group's own enlightened self-interests often involve the use of political power.

In many states, the state teachers' union is one of the most powerful lobby in the state legislature. Having the authority to negotiate contracts at the local level, as is the case everywhere today, puts the local teachers' union in a position of power that equals or even exceeds that of the school board, who are the elected representatives of the people of the community.

Recently, students groups have not been a particularly potent political force in education in this country until relatively. Teachers taught, professors professed, administrators administered, boards of regents set standards for admission and completion, and students could pretty much take it or leave it. The adults made the rules and regulations; students were expected to comply with them. Students could involve themselves in so-called student councils and student governing bodies at the high school and college levels, but these were governing bodies at the high school and college levels, but these were frequently controlled by the institutional establishment.

The governance of institutions was rarely challenged seriously on any issue of substance. All of this began to change dramatically at the college and university levels with the return to the campus of the veterans of World War II. Attending under the Bill of Rights, these men and women invaded the campuses of the country by the hundreds of thousands. They had already had their career plans delayed because of the war, most were several years older than the conventional college student, and most were highly self-directed in terms of their career goals. Also, many were fed up with the bureaucracy and seemingly senseless regulation of military life and were not about to put up with it as civilians.

The students, then for the first time, began putting real pressure on colleges and universities to reform their procedures and to provide a sympathetic ear to the demands of the students. Institutions had little choice, really, for they could not survive without the income these potential students could provide. Thus the World War II veterans of the period between 1946 and 1956 paved the way for student involvement in educational decision making at the college and university levels. Institutions had been accommodating student demands for nearly two decades before the extreme student activism of the 1960s and, therefore, should have been able to cope with it. Nonetheless, when student activism reached its peak in the late 1960s, these institutions found themselves in complete disarray and, in some instances, became practically ungovernable.

These developments had the effect of dramatising the political power of student groups in campus decision making. As a result, students have become active participants in the processes of decision making on college and university campuses. There is an awareness of the rights of students and a sensitivity to the need to involve students in decision making. Today we find student members on many committees that deal with the academic and institutional affairs not only on college and university campuses but in secondary schools as well.

Business and corporate interests organise themselves around such organisations as the chamber of commerce, Kiwanis Club, the Rotary Club, and the local taxpayers' association. These men and women are likely to have many social contacts at local private clubs, at business luncheons, at community affairs, and even in each other's homes. The business and corporate interests have been stereotyped as bedrock political, economic, and social conservatism in this country. There is, however, growing evidence that there is not as solid agreement among leaders of these groups that there once was. Everett Ladd Jr. sees these differences as essentially generational and refers to them as representing 'moderate' and 'orthodox' conservatism. Moderate conservatives accept the service state and the governmental intervention that accompanies it. Even though they may inveigh against it at Rotary luncheons and convention banquets, back in the privacy of the boardroom they try to shape its uses, especially in promoting economic development and in advancing the immediate interests of the business middle class. Orthodox conservatives, in contrast, remain profoundly and genuinely uncomfortable with the New Deal state.

As practical politicians, such conservatives usually understand fully well the limits on how far they may reasonably expect to cut back government. But still, their objective is to produce a situation where government taxes less, spends less, and regulates less. Business groups have often been represented as opposed to good education. This is simply not true. Increasingly, the business community is recognising that schools are vitally important to the equality of life in a community.

Public education and religious groups often encounter value conflicts. The issues become political when pressure is brought to bear on the school to do something that it is not now doing or to stop doing something that it is doing. Examples of the first concern is to persuade schools to engage in prayer recitation and/or Bible reading or to teach the theory of creation. Examples of the second concern are pressures to get schools to stop using enquiry procedures, and so refrain from including sex education or evolution in the curriculum.

The extent to which religious groups are powerful political agents depends largely on the community. In those places where the entire community is of a single religious persuasion, the religious influence can be, and usually is, very strong. If a community is heterogeneous in its religious composition, no one group is powerful enough to dominate the public school decision making.

By the late 1960s, hardly any school decision could be made without taking into account the effect it might have on the minority and ethnic component of the community. These groups gained considerable strength as a result of several favourable judicial decisions and because of federal and state legislation. Although these groups are not as powerful politically as they may have been a decade ago, they will wield formidable political power.

For instance, political analysts suggest that the black vote was responsible for the election of President Carter in 1976. The political and social climate of the period between 1960 and 1970 encouraged self-determination for minority and ethnic groups. Guidelines for federal projects not only encouraged the involvement of minorities and ethnics but required it. The Civil Right Act of 1964 opened many doors to opportunity for these groups and enhanced their political strength. The formation of human rights commissions in states, affirmative action programmes, and a growing public awareness of discrimination also strengthened minority and ethnic groups politically.

Legal Pressure: The courts have been heavily involved in educational decision making in this country. This, of course, changed with the landmark decision of Brown. The resolution of the Brown case in favour of the litigants not only outlawed the separate but equal principle, but it also ushered in a whole new era of the involvement of the courts in educational policy information. This came about, in part, because of changes in judicial procedure that made "class action" suits easier to file a class action lawsuit means that the outcome of the case will affect not only the individual litigant but also all others who are similarly grieved or affected. Thus, a class action lawsuit has the effect of establishing policy.

David L. Kirp has noted that this relaxation of the procedures in filing class action lawsuits, along with the expansion of the substantive meaning of the Equal Protection Clause of the Fourteenth Amendment were instrumental in the judicial system's becoming a vehicle for efforts to secure distributive justice.

As was explained in for a period of nearly twenty years following the Brown case, one case after another found its way into the counts. Clearly, this was done in order to achieve justicc when local institutions were unable or unwilling to provide it. One might say that the unresponsiveness of traditional institutions to human needs encourages judicial involvement in policy formation for public education. But this development was greatly slowed by the US. Supreme Court decision in the Rodriguez case in 1973 and the Bakke decision in 1978. The judicial success in recent years has not been as predictable as it once was, making potential litigants are wary in becoming involved in a lengthy and costly lawsuit that might eventually have adverse class action consequences.

Changes in school funding practices, with a greater share of the costs of education being paid by the state, coupled with citizens' concern over student achievement, have given ascendance to state legislatures as powerful political forces in educational decision making. Another reason for the growing potency of the state legislature is the increased strength of other politically influential groups. For example, the local school authorities can no longer deal effectively with the powerful teachers' unions, business groups, and taxpayers' associations.

These local authorities cry for help from the state legislatures in setting down policies having to do with contract negotiations, funding, student rights, tax rates, and many other complex issues. The same forces that have reduced the influence of the judicial system on educational policy formation have increased the role of the state legislatures. There is little doubt that the state legislature today is the most powerful political body affecting educational decisions.

Increasingly, political groups are recognising this and are taking their problems directly to the state legislature. This applies no matter what the problem might be – curriculum, school finance, personnel, affirmative action, minority group education, and so on. Local political units are not able to deal with these complex problems, and the federal government either refuses to or cannot legally become involved. Consequently, it is the state legislature that has become the educational unit of first and last resort in resolving educational issues.

Social Categories

Naradsmriti, written after *Manusmriti,* permits *niyoga,* gambling as an amusement for women as it was for men, widow-remarriage, etc. Similarly, there are several new features in the *Brahaspati Smriti* and *Katyayana Smriti.* Brahaspati argues that when wife is half portion of her husband when he is alive, how anyone else can have a claim over husband's property when he is dead. There cannot be anything more favourable to womanhood and more democratic in character than the Indian Shiva concept of Godhead itself being represented as a union of half-female and half-male, called Ardhnarishwara. Brihatsamhita, written by Varahamihira in the sixth century AD declares,, "Tell me truly what faults attributed to women have not been practised by men. Women are superior to men who heap abuse on women who are pure and blameless". One may say that the Hindu view of woman has been vacillating between idealising and down-grading her, or one can also call it as varying with different lawmakers and evolving with time and circumstances.

Different Social Systems

A brief view of woman's status in ancient times in different societies, as described by Watermarck, would help us to arrive at considered judgement in regard to the status of Indian women.

"Among the East African Wanika a woman is a toy, a slave in the very wont sense; indeed she is as though she were a mere brute.

Around 479 BC the Chinese philosopher Confucius wrote, "------- (A woman) is subject to the rule of three obediences. When young, she must obey her father and elder brother, when married, she must obey her husband; when her husband is dead, she must obey her son".

A Chinese sayings describes, "Brothers are like hands and feet. A wife is like one's clothes. When clothes are worn out, we can substitute those that are new. When hands and feet are cut off, it is difficult to obtain a substitute for them." Another saying has it, "The best girls are not equal to the worst boys".

Christian religion enjoins a husband to love his wife as his body, to do honour to her as unto the weaker vessel. However, man is not of woman; but the woman of the man. Neither was man created for woman; but the woman for man.

Islam does not give equal status to women although it treats all men equal. The Prophet is reported to have said, "I have not left any calamity more hurtful to man than woman – assembly of women, give alms, although it be your gold and silver ornaments, for verily ye are mostly of Hell on the Day of Resurrection."

Indicators of Status: It may be difficult to predict about the status of women on the basis of certain customs prevailing in the groups to which these women belong, yet it may be useful and interesting to allude to some of them.

"In North-West-Central Queensland, the women, on special occasions, are allowed to inflict punishment upon men; at a certain stage of the initiation ceremony each woman can exercise her right of punishing any man, who may have ill-treated, abused, or hammered her, and for whom she may have waited months or perhaps years to chastise."

Among the Bakongo, a man would be much ridiculed by the women themselves, if he wanted to help them in their work in the field. Sometimes, agriculture is supposed to be dependent for success on a magic quality in women, intimately connected with child-bearing."

In Abyssinia, "it is infamy for a man to go to market to buy anything. He cannot carry water or bake bread; but he must wash clothes belonging to both sexes, and, in this function, the woman cannot help him."

Notwithstanding the customs and rituals mentioned above, there is hardly any room for doubt that in ancient times women in India enjoyed better status than their counterparts did in other societies.

Downfall in Status: Some of the causes leading to the lowering of social status of Indian women, through the corridors of time, are as follows:

1. Even in the Vedas, there are prayers for the replacement of daughter by a son in the mother's womb, for it came to be held that parents attained salvation only through oblation by the son.
2. Non Aryan females of low social classes were admitted into Aryan homes as wives and concubines. Aryans did not consider them fit for a high social status and gradually this led to the downfall in the status of all women.
3. Polygamy contributed to lowering the status of women.
4. Economic dependence of women on men assigned the former an inferior place.

Improvement in Status: With the spread of education and new ideas, the status of women has undergone a considerable change in many parts of the world. The movement to this end started during the Second World War which pulled out women from homes to manage cranes, shop-floors and other services. In the *Scary of Civilization,* Will Durant says (p.131), "Until 1900 or so woman had no right which a man was legally bound to respect. The emancipation of woman was an incident of the Industrial Revolution. They (women) were cheaper labour than men; the employer preferred them as employees to more costly and rebellious males. A century ago in England, man found it hard to get work; but placards invited them to send their wives and children to the factory gate.

Woman's charm and grace is even now employed to promote business. Therefore what we interpret as an exercise for woman's freedom and equality was, to begin with, altogether a different thing.

In India, the freedom struggle, in which women participated in large numbers, as well as influence of the West, contributed greatly in ameliorating the status of women.

To some extent the welfare programmes, but largely the developmental activities-dissemination of education, training women for different vocations, etc., will have an abiding influence in improving the status of women.

Exploited Ones: In India, the status of women varies from class to class. Generally speaking, in the upper strata of society, where people are better educated and there are no economic problems, women have considerable freedom and aspire to be equal with men. Among the working-class, tribals and agrarian societies, women are as good wage-earners as men, therefore they are, perhaps all over the globe as much privileged or handicapped as men and this is the groups with the largest numerical strength in India. In the middle class, the position is different. A Bengali young widow may not be allowed to eat her staple food fish; but there is no such restriction for a widower. Among Muslims, man may divorce his wife; but the wife can not divorce her husband. If a girl is raped, the society is unwilling to accept her for marriage; but if a boy had a pre-marital or extra-marital sex experience, the society compromises with the situation more easily.

While all this may be true, the other side of the coin should also not be lost sight of. One of the post-graduate teacher respondents to the questionnaire wrote, "Man is a victim of exploitation (by women). Cruelty and affliction is no longer a male sport. Education, freedom, economic independence have brought about a great change in the attitude of women and their treatment of men. There are many ways in which women can disrupt matrimonial life like bringing up false charges such as harassment for dowry by the in-laws, ill-treatment and torture from the husbands, etc. Other methods could be nagging the husband, showing indifference to him and cruelty to children'.

There exists an all-India organisation 'Akhil Bhartiya Patni Atayachar Virodhi Morcha' (All India Front for Prevention of Cruelties by Wives) which is demanding adoption of suitable laws for the protection of men from the atrocities of women. The Delhi Chapter of this organisation, in its meeting in Tis Hazari Courts in October, 1988 resolved, "life has treated them (men) badly and there should be a law to protect hapless husbands from the wiles of women."

There was a statement in the questionnaire for students, which touched upon economic and cultural justice (and in a wider sense on social justice as well). It runs as follows:

> "Reshma and Sunderrajan were debating continuance of reservations for Scheduled castes and Scheduled Tribes in engineering and medical colleges. Reshma supported this practice whereas Sunderrajan was against it. What do you think would be the right course for future?

1. Reservations of seats in these institutions should continue;
   Or
2. Reservations should be on merit alone. (S-8)

Reservation of Seats

Sl. No.	*Group*	*Persons agreeing option (a)*	*option (b)*	*Un-decided*	*Persons option (a)*	*option (b)*
1.	School Students (M)	9	60	1	-	-
	(70)	(12.9)	(85.7)	(1.4)		
2.	School Students (F)	10	75	1		
	(86)	(11.6)	(87.2)	(1.2)		
3.	College Students (M)	5	29	-	-	-
	(34)	(14.7)	(85.3)	-	-	-
4.	African College	6	1	-	-	-
	Students (M) (7)	(85.7)	(14.3)	-	-	-
5.	College students (F)	1	66	1	3	-
	(70)	(1.4)	(94.4)	(1.4)	(2.8)	-
6.	Total Population	31	231	3	2	-
	of Students (267)	(11.6)	(86.5)	(1.1)	(0.8)	-

A glance over the above table would reveal that students opposed the idea of reservation of seats in professional institutions. Lack of support to the idea of reservations can emerge from several considerations. Firstly, institutions for professional training, in a way, pave a way to various job opportunities. Jobs are further related to the economic well-being of the country and managing positions of public responsibility in both case great care is needed. For example, if one is having a health problem, one would like to get treatment from the best possible medical practitioner, rather than one who owes allegiance to any particular caste or religion irrespective of his competence.

This does not mean that persons belonging to SCs and STs have low merit. The Chairman of the Drafting Committee of the Indian Constitution, Dr. B.R. Ambedkar, was a Scheduled Caste Mahar. The writers of the *Ramayana* and *Mahabharata* also came from the Scheduled Castes. Balmiki was originally a DALIT and Vyas was born of Satyawati who was a daughter of a boat-man and a DALIT. Merit is no monopoly of any particular group.

However, it is necessary that adequate provisions should be made for those sections of society who have been educationally left behind so as to enable them to make up the lee-way as quickly as possible. Many among the Scheduled Castes and the Scheduled Tribes feel that the provisions of reservations has not benefitted them as much as the element of discrimination against them by higher castes has harmed them. Therefore, today there is a great need for removing discrimination against them, wherever it exists. Lastly, some people are of the opinion that reservation policy should have been founded on economic backwardness and not on the basis of castes.

Framework of Society

Society can be looked upon as a process; as a series of interactions between human beings; each person stimulating another person and responding to the stimuli from the other person. No social life is possible without such interactions. This is why communication is basic to social life. Communication is interaction in terms of a stimulus, a gesture, or a word which one person produces in response to another stimulus, gesture, or a word from the other person. The mother-child relationship is a typical example of the interaction process.

The care of the child, the upbringing and socialisation of the child is based on the interaction processes between the mother and the child. In fact, the personality of the child is largely determined by the socialisation process in this early relationship with the parents and others at home and later on with the teacher and the peer group at school. But one should not imagine that the personality of the child is wholly or completely determined by this interaction process, in the sense that the child is a mere replica of the parents. Each child starts with his own needs and psychoneural bases for emotional experiences and cognitive orientations, and develops his own personality. The interactions with others may lead to cooperation as well as to conflict. While cooperation is the process of working together for commonly accepted ends, conflicts arise when there is a struggle between rivals for the same goods and services or for recognition. Besides, there is the process of competition between the two interacting persons or groups when they strive to attain the same goods or services or recognitions. Thus, interactions lead to uniformities as well as diversities, so that while there are many similarities between the individuals in a group, whether it is a family group or a neighbourhood group or a national group, there are also many diversities. In fact, as our experience makes it obvious, each member of the family and each member in the classroom or of a football team is a unique personality. In the same way each group is also unique. Just as no two individuals in a family are alike, no two families are alike. As we observe larger groups, these diversities and unique features become more obvious. Thus, the whole human society and each group in the society can be viewed as the manifestation of the social processes between the interacting members.

The concept of social process is closely tied up with the evolutionary thought that society is a developing organism, that society has passed through evolutionary changes in the sense in which Charles Darwin (1809-82), the British biologist showed that species of organisms have evolved. There is no doubt that the concept of evolution has been very helpful to understand the development of various kinds of organisms. In the biological field, the term evolution indicates the emergence of new structural organisations. There is certainly nothing analogical to this in the social field. While the structure of the tribal society in the northeastern frontier of India is totally different from the structure of the society in a metropolis like Bombay, by no stretch of imagination can this difference be equated with the differences, either anatomical or behavioural, between the fly and the rat or between the rat and the man. However, though the evolutionary theory of society may be discredited empirically as well as theoretically, there is no need to discredit the concept of social process.

The concept of structure indicates an ordered arrangement of parts. The parts may vary but the structure is persistent. The interactions or processes are accomplished with this persistent framework of the given social structure. The advantage of the concept of structure is that it helps in the analysis of social events at a given time. This is why short-term empirical studies can be more easily made using the concept of structure.

However, when a social system is considered over a period of time, it is necessary to look at the origin and growth of the system. Though Parsons is a structuralist, he relies on the concept of social process in order to account for social change. The structure itself may change through time. As Parsons writes, "A social structure is a natural persistent system. It maintains its continuity despite internal changes from moment to moment, year to year, just as a living organism remains the same living organism in spite of metabolic changes; it is dynamic" (Parsons, 1965).

It is generally argued that in studying 'structure', we study essentially the interrelation or arrangements of parts in some total entity. The components of social structure are the human beings; the structure here is the arrangement of persons in relationship which are institutionally defined and regulated (Radcliffe Brown, 1957). Further, social structure also consists of ideas regarding the distribution of power between the persons or the groups at a given time. Thus, as Nadel puts it, "social structure is a conceptual tool; it is an abstraction indicating the network of relationship between the persons who have different roles to play" (Nadel, 1965). The basic social data consist of the interactions between persons. The term 'role' is used to refer to the regular repetition of interactions over a period of time. It is clear that in any social organisation like an administrative office or factory producing goods, there is a social structure, in which various persons in various positions, from the head of the organisation to the peon and the sweeper, play the roles assigned to each one.

The persons playing the roles may change from time to time but the roles played will be fairly constant. As Smelser (1970) puts it, "Social structure can be defined as identifiable patterns of roles that are organised primarily around the fulfilment of some social function or activity." Another aspect of the problem is that all social behaviour is subject to certain kinds of social regulations, sanctions, norms and value orientations. Thus, the main ingredients of social life are the interactions between persons who are in role relations to one another and whose interactions are governed by sanctions, norms and values.

Modern Line

The very idea of human society presupposes order. The interactions among the actors in the system are reciprocal and interdependent. When these interactions are repeated and persist over a period of time there is an order in the social relations. The term social order refers to the existence of *restraint,* the inhibition of impulse, or more specifically, to the control of violence in social life. Secondly, as noted above, social order refers to an element of *reciprocity* or mutuality in social life. Next, it refers to an element of predictability in social life; that is, there are certain expectations in interactions. The fulfilments of these expectations is possible when there is some *consistency* in the interactions. Finally, social order implies *persistence.* As Cohen (1968) puts it, all these meanings of the term social order are logically and empirically related. Thus, any society is the organisation of human relationships. This constitutes the social structure.

The ordering of human lives in terms of a matrix of social expectations comprises part, or perhaps all, of the distinctively human qualities of behaviour (Moore, 1967). Social order is based on learning; individuals learn to internalise the restraints during the process of socialisation in early childhood. This implies the existence of rules or norms. This is how the society patterns and regulates individual behaviour (Kuppuswamy, 1973). The ancient Indians identified these norms which hold together the individuals in the society as the *dharma.* Literally *dharma* is that which holds a thing together, makes it what it is and prevents it from breaking up and changing into something else. The term *dharma* is derived from the root *dhri* which means to sustain, uphold, to hold together.

It is obvious that without norms, social relations would be haphazard, chaotic and dangerous. It is the norms which give order, stability and predictability to social life. Norms are at the basis of social structure. As Durkheim pointed out, a situation of complete normlessness or *anomie* would be intolerable. No normless or anomie society could endure. As anarchy is the contradiction of government, so anomy is the contradiction of society. No society can exist when there are no norms to govern the social relations of the members. This is why norms constitute one source and locus of the order that society exhibits. In a total absence of norms men would be unable to get together for any purpose whatever; no civil society would ever be possible.

According to Durkheim, the social solidarity of mankind in the past was merely *mechanical* based on kinship, friendship, etc., while in the modern world social solidarity is increasingly *organic* based on division of labour. However, he showed that division of labour sometimes creates the very negation of solidarity. He wrote "If the division of labour does not produce solidarity ... it is because the relations of the organism are not regulated, because they are in a state of *anomie*. Since a body of rules in the definite form which spontaneously established relations between social functions have taken in the course of time, we can say, *a priori*, that the state of *anomie* is impossible wherever solidary organs are in contact or sufficiently prolonged". Thus *anomie* is the product of urbanisation and industrialisation.

It is the compliance with the social norm which promotes social order. Sumner (1907) showed the distinctions between the various type of restraint exercised by the society on individuals; *folkways* imply that one is normally "expected" to comply with the norms; *mores* imply that one "ought" to behave in compliance with the norms; *laws* require that one "must" comply with them. But compliance with social norms is not always or necessarily based on external restraints. In a properly socialised individual who has developed his own ideals of behaviour, compliance is due to internal restraints which may be based on guilt feelings or on a spontaneity which arises from the free and voluntary acceptance of certain ideals and rules of conduct.

Merton (1957) has analysed the various sources of social deviation. He has shown that conforming behaviour requires the acceptance of both the norms (institutional means) and the values or the cultural goals. Deviant behaviour may arise from adhering to goals while departing from approved means as in *innovation;* or from rejecting values while complying with standard rules, as in *ritualism;* or from rejecting both values and norms as in what he calls *retreatism;* he adds *rebellion* which rejects the extant values and norms and the espousal of others.

According to Merton *conformity,* represents the adaptation in which both the cultural goals and the norms (the institutional means) are accepted. Most people in a given society, occupying any position, are conformists. Their commitment both to ends and to norms constitute the chief components of the stable society. As Merton puts it, "It is, in fact, only because behaviour is typically oriented towards the basic values of the society that what may speak of a human aggregate as comprising a society."

Innovation is the rejection of institutional norms while accepting the culturally prescribed goals. He gives the illustration of those who accept accumulation of wealth as a success goal but use various means unapproved as institutional means to reach that goal. He also gives as illustrations the "white collar" crimes among the members of the upper classes and various forms of delinquency, crime and racketeering among the members of the lower classes; crime and delinquency, thus, are illustrations of the substitution of "proscribed" norms in the place of "prescribed" norms.

Ritualism is a mode of adaptation by which the cultural goals are rejected while the institutional norms are accepted. He looks upon, such deviation as "a *private* escape from the dangers and frustrations which seem to them (the lower-middle class) inherent in the competition for major cultural goals by abandoning these goals and clinging all the more closely to the safe routines and institutionalised norms." The ritualist slavishly follows the rules not because of over-identification with them but "from a lack of security in important social relationships in the organisation". He is a conformist to the institutional norms though he rejects the culturally prescribed goals and replaces them by goals more possible for him to achieve. In passing, it may be mentioned that many social problems arise in contemporary Indian situation because of this ritualism among the higher caste persons, particularly in the rural areas, often even in the urban areas.

Retreatism, according to Merton, is a "privatised rather than collective mode of adaptation" of the socially disinherited. They reject the cultural goals as well as the social norms, the institutional means. In this category of deviants Merton includes the vagrants, the tramps, the vagabonds, the chronic drunkards and the drug addicts. He also includes in this category the persons who have suffered an abrupt break in the familiar normative framework and established social relations in their lives like the woman who is overwhelmed by sudden widowhood, or the man who withdraws from all social relations because of some disastrous experience. In such deviants there is a failure to look upcn anything as valuable, anything as worthy of effort; they are cynical; they are disenchanted with the cultural goals; they do not care to.observe the norms.

Rebellion is the rejection of the prevailing goals and substituting them with new goals and also the rejection of prevailing norms and substituting in their place new norms. This presupposes alienation from the generally accepted goals and standards and an attempt to bring about a new social structure. This rebellion "involves a genuine transvaluation, where the direct or vicarious experience of frustration leads to full denunciation of previously prized values". According to Merton there are two factors which contribute to adaptation by rebellion. One is the pressure for achievement coupled with a realisation of the existing restriction of opportunity. The other is ambivalent or conflicting norms resulting from an admittance of open class and casteless norms in the society.

Innovation, ritualism, retreatism and rebellion are individual valuation of various aspects of the goal-norm complex. Merton looks upon them as a typical to the society as a whole, though occurring with sufficient frequency in significant numbers of individuals. This categorisation and analysis, according to Merton, also helps to understand the phenomenon of *anomie* as noted by Durkheim. He further notes that these non-conforming adaptations are not "rationally calculated and utilitarian"; on the contrary, since they arise out of pressure and frustration, a degree of irrationality might be expected.

While crime as part of innovation, ritualism and retreatism and even rebellion of some kinds lead to social disorganisation it cannot be said that all deviation, particularly innovation in technical and organisational field and rebellion against social injustice and exploitation can be looked upon as leading to social disorganisation.

Three terms are in usage, namely social order, social stability and persistence. It is true that the opposite of the first two terms are social disorder and social instability. This is why in general all societies are against innovation and rebellion because they envisage such movements to lead to social disorder and social instability and all the consequences of normlessness and violence associated with them. As an illustration of this fear may be given the instance of resentment against the Hindu Code Bill in early fifties among the Hindus. It was widely believed that the right to marry another wife when one wife is living was sanctioned by religion. There is a similar resentment now among the Muslims to bring about uniform personal law as required by Art. 44 of the Indian Constitution.

However, the opposite of persistence is change and even the most radical changes, as noted earlier, do not lead to social disorganisation, though they may lead temporarily to social instability: the successful radicalist is as eager to preserve social order and stability as the most conservative person. The radicalist is actively and fervently interested in social order and social stability since it is in such a social situation that he can work out the radical programmes of his ideology as exemplified by the Russian revolution.

Social change takes place when there is a structural change in society. With technological and institutional innovations there are changes in economy as well as in the aspirations of the people. When cultural changes as well as changes in social norms take place in response to the new situation, social change may take place smoothly. Otherwise there may be rebellion and social disorder.

As Talcott Parsons has said social systems have to be conceived as *open systems* engaged in complicated processes of interchange with the environing systems. Internally, the social system has to be conceived as differentiated and segmented into a plurality of subsystems, each of which must be treated analytically as an open system interchanging with the environing subsystems within the larger system. Thus there is a "strain" within the system. If the strain becomes too intense, the mechanisms of control and restraint will not be able to ensure conformity; as a result there may be the breakdown of the structure. As Parsons puts it, the strain may be relieved either by restoring full conformity with the normative expectations or by a change in the structure itself. He defines structural change "as alteration in the normative culture defining the expectations governing that relation". As regards the problem of identity of the changing social system he asserts "It follows that the crucial focus of the problem of change lies in the stability of the value system. A change in structure of a social system is a change in its normative culture." The most important factors favouring structural change are adequate mechanisms to overcome the vested interests which generate resistances of institutionalised structural patterns (Parsons, 1961).

The only opposite of social order is not disorder; similarly the only opposite of social stability is not instability; change is the opposite of persistence. Social change may take place with the least disorder and instability or after considerable disorder and instability. But change can only take place in something persisting; further with change, order will be restored and stability will be established to implement the change programme even after a violent revolution. Social change involves an alteration in social structure, certainly with new social norms being established, probably also with new social values. Analysis of the various cultural systems show, however, that the new elite can trace elements of the new social order to the values cherished in the old order. In general, it may be stated that there are greater possibilities of generating new social norms and new social institutions than of generating new social values. This, at any rate, is true of Indian culture in which the new social value enshrined in the Indian Constitution can be traced to the Upanishadic age and to the epic age of the ancient times, particularly to the Gita, Shantiparva, etc.

Many a Change

A social structure is a nexus of present relationships. It endures because it is maintained by the members of the social system in the present. As MacIver and Page (1952) put it, social structure is a web that exists only as it is newly spun. It appears to persist because the attitudes and interests of the members persist. It is the members who maintain its continuous existence. "The most sacrosanct and seeming permanent institutions exist by no other right and in no other strength than that which they derive from the social beings who feel and think and act in accord with them," Social stability and social change are both the features of any social system. It is generally assumed that the tribal society and the village society are stationary.

This is because we know hardly anything of their past it is also because there are hardly any technological changes; the tools as well as the techniques used are practically unchanged, transmitted as they are from father to son; they live in relative isolation; the various subgroups within the small group, accept the social structure as something permanent and unchangeable; probably it is looked upon as sacred and it will be a sin even to think of any change. But changes do take place as the anthropologists who have revisited these groups have shown, though the members do not perceive any change whatever probably because they do not look upon them as perceptibly significant.

As noted above a close study of Indian cultural, religious and social institutions clearly show profound changes which have taken place from time to time though the members of the group feel honestly that nothing has really changed. As Buddha pointed out long ago, persistence and change are both characteristic of not only objects but of individuals and social systems. Both persistence and change are problems for study in sociology. The social system changes in response to changing conditions within and without, internal as well as external.

Change in Society

Towards the end of the 19th century and at the beginning of the 20th c, sociologists and anthropologists were preoccupied with the problem of how societies evolved, from their original primitive state. The concept of evolution formulated by Charles Darwin in the biological field was applied to the social field. Also the great progress made by the Western European nations in the field of colonialism in 17th and 18th centuries, in the field of political organisation with the development of democratic institutions, and in the field of industrialisation and economic organisation in the 19th century, stimulated the social thinkers to account for the onset of these changes. But the great devastation which took place in Europe as a result of the First World War (1914-1918) and the tremendous communist revolution in Russia (1917) made the social scientists to concentrate more on the problems of social structure and maintenance of order. However after the Second World War (1939-1945) and with the emergence of new states since 1947 when India obtained independence, there is again a preoccupation with the problems of social change, and social development. When the United Nations Organisation was formed in 1945 there were only 51 member states; by 1960 there were 100 member states and today there are 191 member states. The former colonies in Asia and Africa have now become independent states; they have accepted modern forms of political organisation; their aim is to adopt the modern economic organisation by industrialising their economies so that the standard of living of the people is raised. These great changes have now induced the sociologists to be engaged in the study of social change.

The Indian sociologists are greatly interested in the problem of social change since the Indian society has now taken up the task of changing itself from an agricultural society to an industrial society, from a colonial society with Emperors and monarchs to a republican society, from a society based on caste and class to a society which aims to be casteless and classless with equality of opportunity to every citizen guaranteed by the Indian Constitution adopted in 1950.

In other words, the urgent problem of contemporary social situation in India is the transformation of the individual from a member of a tribe or a village or a caste or a creed or a language group to a citizen of India. This transformation of man and society has been the central and quite dominant concern of sociology right from the time when it emerged as an independent branch of learning. One of the chief aims of sociology has been to study the conditions under which the disruption of the old traditional order of society occurred and the conditions which lead to the explosive emergence of a new individualism. During the nineteenth century, as noted above, the main aim of sociology was to understand this rapid revolutionary transformation of society. Comte (1798-1857). J.S.Mill (1806-1873), Spencer (1820-1903) and Marx (1818-1883), in their several ways provided an analysis of this transformation of society. Tonnies (1855-1936) in his book Community and Association constructed two clear 'types' of society, the living community (the gemeinschaft) in which all the members were bound together in a shared order of kinship and neighbourhood, living and working within a known and loved territory with clear and well-understood rules, values and social arrangements and the complex network of formally and rationally devised associations (the *Gesellschaft)* of which men were members only in a contractual sense, in which the relationship rested entirely on means and calculations, in which both men and principles became impersonal.

This transformation of men and social relationship is necessitated by the new social, political and economic organisations. Political independence and the formation of one central government, urbanisation on a vast scale with the development of the great metropolitan cities, the planned industrialisation and the development of rapid modes of transportation and communication, the development of huge banking organisations with branches in most of the villages and all the towns and cities, the establishment of primary schools in practically all the villages, all these developments since independence have necessitated a transformation of the society. So it is the task of the social scientist to study how the transformation is taking place and analyse the nature of the resistances which are preventing or which are slowing down the speed of the transformation.

Process in Continuation

However, as Becker (1957) pointed out, any social system may be viewed as occupying some position along a continuum varying from maximum reluctance to change to maximum readiness to change. He even goes to the extent asserting that a society may run the risk of extinguishing itself by preventing change or, on the contrary, may extinguish itself in pursuit of change. In other words, both maximum reluctance to change and maximum readiness to change are suicidal; no society can survive under either condition. Survival depends upon some flexible approach which takes into account the situation which arises when factors inducing or necessitating change are operating. Such change- inducing factors may arise from within or from without.

In a broad way it may be asserted on the basis of the anthropological and sociological data available, that the more primitive societies are more reluctant to change while the more modern societies are more ready to change. To put it in another way, the simpler societies are more reluctant to change while the more complex societies are more ready to change. In fact, the more modern and more complex societies have inbuilt mechanisms, as it were, to change. For example, the more modern societies encourage the growth of science and promote facilities for the application of such new knowledge by establishing special institutions for scientific research and technological progress. Similarly the political institutions in the democratic framework provide for periodical general elections and formation of new governments. The Parliament is empowered to repeal old laws or amend them or bring on down. Public administration provides the framework through which the controlling authorities are transferred from one section to another.

Similarly research in agriculture has been sponsored in the universities of agricultural sciences in various parts of the country, by the Indian Council of Agricultural Research, and the Indian Agricultural Research Institute, New Delhi; besides several special research institutes like Central Rice Research Institute, Cuttack; Indian Institute of Sugar Technology, Kanpur and others. The universities of agricultural sciences as well as the state departments of agriculture have their extension wings which make the knowledge of new methods and the facilities and equipment available to the farmer. High yielding varieties of foodgrains, increased use of chemical fertilizers and the maximum use of minor irrigation facilities were all introduced as a package deal in 1967-68. This led to a great spurt in agricultural production in what is called the "Green Revolution".

In the same way there is what is called R and D, the Research and Development Organisation, in the Defence Services of India to promote and apply scientific research in the Army, Navy, Air Force and Communications. Several institutions are engaged in advancing knowledge as well as technology; training courses are conducted so that the personnel are not only equipped with but also trained to use the new techniques.

In this context, it may be recalled that Gandhiji formulated the principle of satyagraha as a form of protest against the unjust laws or the unjust government and developed the technique of non-cooperation to see that political, economic and social changes can be brought about in a social system by mobilising public opinion and public action. One of the outstanding events of human, social and political history is the fact that the political struggle in India led to the intervention of the British Government to establish, through negotiation, the Constituent Assembly to frame the Indian Constitution and to the attainment of Independence in 1947, which set in motion the liquidation of the colonial form of government in the various countries of Asia and Africa.

One of the established facts of history is that the prediction of Karl Marx regarding the polarisation of classes in an industrial society did not materialise since the state as well as the society in western European nations and in North America brought about changes in the working conditions through reform, legislation and economic reorganisation and by legitimising the trade union movement among the workers.

Thus, the more complex societies have built-in mechanisms in the various institutions to enable them to change themselves in response to the strains and stresses which arise and also in response to the new knowledge and techniques deliberately developed to overcome the strains and stresses:

Long ago, the ancient Indian thinkers tried to discuss the problem of persistence and change. The Upanishads; nearly three thousand years ago, looked upon the *Brahman* as the ultimate reality and asserted that the *atman,* great sayings *(mahavakyas)* of the Upanishads like *tattvamasi* (thou art that), *aham brahma asmi* (I am Brahman). The self as well as the universe are real and permanent, *nitya.*

On the other hand, Buddha (2500 BC) developed the doctrine of momentariness *(Kshana-bhanga vada)* and impermanence *(anityatva)* and asserted that nothing that is, lasts for longer than one instant. While it was commonly assumed that the self not only persists as long as the organism persists, but also that the self survives the body, Buddha declared that this was an error. The self, Buddha said, is not permanent; it is undergoing change almost constantly. Thus, Buddha postulated a changing self, as a protest against a static, permanent self postulated in Upanishads.

But the problem arises as to what is changing; something must exist and persist to change. Buddha explained that what exists is the aggregate, the, *samghata;* he even accepted the doctrine of *karma* and *samsara,* that one has to go through a series of births to perfect oneself to attain *nirvana* the state of non-being. Thus the Buddhists believed both in the doctrine of unenduring self and in the doctrines of rebirth *(punarjanma)* based on the fruits of *karma.* However as Hiriyanna (1932) puts it "The belief in the karma doctrine really presents no new difficulty to Buddhism; for if there can be action without an agent, there can well be transmigration without a transmigrating agent. Further, we have to remember that according to Buddhism there is transmigration or, more precisely, rebirth, not only at the end of this life as in other Indian beliefs, but at every instant. It is not merely when one lamp is lit from another that there is a transmission of light and heat. They are transmitted every moment; only in the former case a new series of flames is started."

The same problems of permanence *(sanatana)* and change *(parinama)* was discussed by the ancient writers on social problems. For example, Manu is aware of the social changes which had taken place nearly two thousand years ago when he wrote his code of laws. He asserted many customs practised by the ancients had to be discarded because of *yuga dharma.* He recognised that each period of time had its own norms of behaviour and judgement and its own set of values. This is why Bhishma explains in Shantiparva (78:32) "Righteousness becomes unrighteousness, and unrighteousness becomes righteousness, according to place *(desha)* and time *(kala).* Such is the power of place and time (in determining the character of human acts)." As Prabhu (1963) writes, the recognition of the role of time and region in the formation of *dharma* is an indication of an awareness that the concept of *dharma* was no rigid concept but a flexible, modifiable and dynamic principle related to the exigencies of time, place and human needs. "And, moreover to speak of *dharma* itself becoming *adharma* and vice versa according to the times and the locale is really to speak of much more than just allowing the possibility of modifications in *dharma;* it is to ask man to be prepared for such complete and radical changes in *dharma* which may even be contradictory to the generally sanctioned *dharma* practices in order to suit the exigencies of the time and the conditions, if the latter necessitate such changes. It is in view of the impossibility of visualising the emergencies of the locale and the times that the *sastrakaras* have not attempted to dwell on the *dharmas* of desa and kala."

Thus, the ancient Indians, nearly three thousand years ago, were aware of the general problem of permanence *vs* change and the specific problem of social change. But the general belief, throughout the millennia, has been that the social system is something god-given and that it was the duty of the political authority to preserve and enforce the *sanatana varnasrama dharma,* the permanent, unchangeable, social structure based on the four *varnas* (castes) and the four *ashramas* (the stages of life) (Ghosal, 1966).

Cohen (1968) has discussed this problem of continuity. He asserts that it is obvious that one cannot conceive of a social structure or social system unless one assumes that social life has continuity. Even when one assumes that all societies are constantly undergoing change, it is important to recognise that what changes is the social structure. As he puts it, this is "another way of saying that the study of social change is the study of the disruption of social persistence and the study of social persistence is the study of social processes which inhibit or fail to produce change". As he puts it elsewhere "No society is persistent without change, nor changing without being persistent". Discussing the problem of partial change and total change in social structure he points out that when the whole system changes, there is no way of identifying it is as the same system. It is clear that even when there is 'total' change as in a revolution, something persists, something has not changed so that it is possible to say that a change has taken place in the given social system.

The conundrum can be clarified by a few illustrations from recent social history. There is no doubt that the Russian Revolution of 1917 was a great revolution in the total structure of the Russian society. But it is clear that the Russian society has endured in spite of the drastic change in the social structure. There is a continuity. Even the big change in the post-Stalin era in 1964 did not "completely" alter the Russian Government or the Russian people any more than the 1917 events. Many aspects of Russian culture and many aspects of Russian personality structure have endured in spite of the drastic changes in the Russian social structure. The same thing can be said of the Chinese Revolution of 1948.

Another illustration can be given regarding some profound changes in culture, which have altered, but not 'completely' changed the Hindu culture. It is a familiar fact that the Vedic Hindus (1950-1000 BC) were completely dominated by ceremonials, the *yogas* and the *yajnas* involving animal sacrifice. With the complexity of the sacrificial ritualism, the doctrine of the four *varnas* and four *asramas* began to take shape.

The *Upanishads* (800-700 BC) represents a spirit different from and even hostile to ritualism. The Upanishadic theory of universe is quite distinct from the theory of universe of the Brahmanas. In fact, one of the clearest onslaughts against sacrificial ceremonial is to be found in the Mundaka Upanishad (I. II. 7) which asserts that one who hopes for real good to accrue from these rites is a fool. In the age of Upanishad *jnana* took the place of *yajna*. Further not only the beliefs in gods and the sacrifices but also the practice of caste system was condemned.

After the Upanishadic age (600-400 BC) the sacrificial rites became prominent once again and there was a greater rigorous insistence on the *varna* doctrine. Thus, the orthodoxy of the priests remained practically unaffected by the Upanishadic philosophy. It is in this period (560-200 BC) that Buddhism and Jainism became powerful emphasising the ethical ideals of the Upanishadic age but condemning the authority of the Vedas, the ascendancy of the priests, the supremacy of sacrificial ritualism and the practice of caste system. It was also in this period that non-violence, *ahimsa*, became most influential and the worship of images in temples started. As Sarma (1956) writes "The cult of the worship of images seems to have arisen spontaneously when the Vedic sacrifices became too elaborate and complicated and the various *Upasanas* recommended in the Upanishads came to be substituted in their place" (p. 10).

In the Epic Age (200 BC to AD 300) with the fall of the Mauryan Empire, the prestige of Buddhism suffered and there was a revival of Hinduism. The Ashvamedha sacrifice was performed by Pushyamitra, the Brahmin general, who killed the last Mauryan Emperor and founded the Sunga dynasty. It was in this period that the great epics, Mahabharata and Ramayana took their final shape as didactic books. It was also in this period that the codes of Manu and Yajnavalkya were composed. All these developments brought about a great change in the outlook of the priest class in particular and the people in general. The movement set afoot by Buddha modified the Hindu beliefs and doctrines. As Sarma (1957) writes "The gates of the temple were at last thrown open to the classes. The knowledge which had remained the exclusive possession of a small class was made available for all. Not only that, there was a fusion of the Aryan culture with the Dravidian culture. The gods and goddesses worshipped by the common people were given honoured places in the Hindu Pantheon" (p. 22). The organised sects of Vishnanism, Sivism and Saktaism (the worship of Durga) came into prominence. In this age the sacrificial altar gave place to the temple and image worship. Both the *Yajna* of the early Vedic age and *Jnana* of the Upanishadic age gave place to *Puja* and *Bhakti*. The Hindu scheme of life as expressed in the formula *Dharma-artha-kama-moksha* was fixed and widely taught.

All these new developments in Hinduism are expressed in the Gita. The traditional concepts of *Yajna, Karma, Varna* and *Dharma* are given new interpretation. According to the Gita, *Yajna* does not mean animal sacrifice; it means the pursuit of all activities of life in the spirit of sacrifice. Similarly *karma* in the Gita does not mean ritualistic actions but all human actions having moral or spiritual value. *Varna* is not based on birth but on *guna* (character) and *karma* (moral activities). *Dharma,* according to the Gita, is not simply the caste-duty of popular ethics, but the duty imposed on man by his own ideal of life *(swadharma)*. Another outstanding feature of the Gita is the formulation of *Bhakti,* devotion to God. "Thus the Gita everywhere follows the old tradition, but everywhere extends it in such a way as to recreate it. It retains the old Upanishadic ideal of *jnama,* but balances it with *nishkama karma and bhakti*" (Sarma).

This short description of the various phases in the Hindu religion from the Vedic period (2000 BC to the Epic period AD 300) clearly shows how profound changes in the concepts, beliefs as well as practices did not affect the continuity of Hindu religion. As it is well known, the Gita is one of the most important books in the Hinduism of the 20th century and has led to various commentaries by the leaders of thought and action from Tilak to Gandhi. This shows how changes do not lead to disruption in social and cultural life, however, profound and drastic they may be. The social system persists with all the changes incorporated.

It is like the profound changes, physiological, psychological and social which take place in each individual. When the child is born, he is small and helpless. As he grows up he is highly egocentric. With the further growth, profound changes take place in his body, mind and social behaviour during adolescence. Further changes take place in adulthood when he takes up an occupation, marries and settles down to rear his own family. Finally, further profound changes take place when he retires from his occupation and also from his family life. But throughout he is the same individual, having bodily, mental and social identity. The ancient Indians recognised this and formulated what they called the *ashrama dharma.*

From the foregoing it is clear that the distinction between persistence and change is more a matter of approach rather than something substantive.

There are two well-established types of social enquiry – the *synchronic* and the *diachronic.* The former is concerned with static or simultaneous study of an array of facts in a social system, while the latter is a study of the phases separated in time, a study of the successive changes in a society. It is clear, however, that the two terms should not be understood in a rigid manner since nothing in the sphere of social facts is literally static. The subject-matter of sociology consists of regular, repetitive social behaviour. It is obvious that such a regularity in behaviour is only visible over a period of time. From this point of view, the synchronic approach limits itself to the study of simultaneous as well as continuous states of affairs in a society over a short period of time. By contrast, the diachronic approach is concerned with broadly separated time phases in a society. It is this approach which helps the study of social change, namely, the change in social structure (Radcliffe Brown, 1957).

Those who are interested in the problems of social stability and who consider that the main task of sociology is to explain social stability adopt the synchronic approach. They look upon social situations as though they were static or endlessly repetitive; as a result they study the problem of social structure as it exists at the given time and the social forces which are conducive to the maintenance of the system. They show that human acts, groups, rules of conduct and goals or values are interrelated. As Moore (1961) puts it "The inter-relationship constitute the major data for meaningful description or analysis of social systems. By their predictability, their persistence, they constitute a major source of order, of reliability in social affairs...." He then proceeds to observe "Besides the orderly persistence of interrelated patterns of action, there are also ample sources of disorder in social systems, of tensions and conflicts, of intrinsic sources and paths of change.... If an equilibrium model is taken at all seriously, these circumstances are downright embarrassing." This is the main difficulty of the synchronic approach to the study of social systems. With emphasis on social order and social stability, every conflict and tension is looked upon as a source of disturbance, a source of disorder and confusion. Cultural and social change is looked upon as a separate field of enquiry involving separate methods of study and separate models.

But there are others who seek out situations furnishing evidence of development or other forms of change. As Nadel (1951) puts it, "In describing social phenomena we can choose whether to consider some roughly contemporaneous phase or several such phases stretching back in time; in the latter case, we might well be describing also processes, developments, that is, phenomena of change. But in explaining social phenomena we have only one method at our disposal, namely, that of examining their interdependence; and this, in turn, is visible most clearly when change sets in, that is when some variable in one phenomenon can be seen to promote concomitant variation in others." It is obvious that diachronic approach is more helpful to study the social system from this point of view. Nadel also makes the point noted above that at one end there are societies which are rapidly changing and interdependent and at the other end there are the societies which are self-contained. But it must be acknowledged that with the rapid growth in the means of transportation and the means of communication and with their immense spread, it is hardly possible for any social system to be so self-contained, at any rate in India.

Next, there is the other problem of planned social change; in all contemporary states and certainly in the states which have emerged recently, five-year plans have been launched. With political independence, the people in these new states are eager to have higher standards of living and greater access to modern developments. It is necessary to study the extent to which such planned changes have been effective why there are resistances to them.

Social Empowerment

Chance for Equality

Notably, the word 'gender' means much more than 'sex'. The gender of a man is masculine and that of a woman feminine. Neither a man nor a woman is sex alone or a biological species. So, the concept of gender may be said to be more inclusive than that of sex. But this distinction as such says nothing as to which of the two is inferior or superior. A bigger circle is only bigger than- not necessarily superior to — the smaller one, which it may include. In other words, the word gender in my view is a value-free concept. Differences of value arise only when we take genders in relation to their functions in society and, what is more important, when our way of looking at the matter is merely external. The point may be brought out as follows:

When female child becomes a wife, she acquires the functions of a mother, similarly a male child becomes a father. Now the functions of a mother are mostly confined to activities which quietly take place within the house. The father, on the other hand, earns a living by working in the outer world. What takes place in the open is noticed easily. What happens at home does not strike the public eye. So by the average man, whose way of looking is confined to the externals — that is, whose *drishti* is *bahirmukhi* —

the male is taken to be superior to the female. Those who are careful enough to take a comprehensive view of the human life, attach as much value to the mother's activities of producing and nursing children and keeping a family together as to the bread winners outdoor activity of earning a living. It is really a defect in our ways of looking at things, and not the fact of gender as such, which is at the root of prevailing bias against women.

However, in the contemporary feminist literature, gender is not a value-free concept. It is a value-loaded term. It has acquired new dimensions and greater significance. It now refers to the social institutionalisation of sexual difference . It aims at exposing the present masculinist hegemony in the name of natural sexual differences, and also at uncovering the male connotations of the existing vocabulary of reason, morality, autonomy, justice and history. It, therefore, aims at suggesting an alternative epistemology and methodology, which can uncover the present gender bias, and may reflect more accurately the experiences and needs of all human beings. Feminists assert that any discrimination based on the basis of sexual differences is unjust; the body differences do not warrant such discriminatory differentiation; and, that they are only socially produced. Catharine A Mackinnon says:

> "Our issue is not the gender difference, but the difference gender makes, the social meaning imposed upon our bodies - what it means to be a woman or man is a social process and, as such, a subject to change. Feminists do not seek sameness with men. We more criticise what men have made of themselves and the world that we, too, inhabit. We do not seek dominance over men. To us, it is a male notion that power means someone must dominate. We seek a transformation in the terms and conditions of power itself."

The Legacy

It would be relevant here to discuss how political theorists in the past have neglected gender. John Locke defines political power as distinct from the power relations operating within the household. Rousseau and Hegel have clearly contrasted the two spheres and have justified this contrast in legitimising male rule in the domestic sphere. Locke has categorically mentioned that when "women hold the helm of government, the state is at once in jeopardy". Rousseau believes that women pose a permanent threat to political order. The natural morality of women fits them only for the 'natural society' of domestic life. He argues in Politics and the Arts, "even if it could be denied that a special sentiment of chasteness was natural to women, would it be less true that in society ... they ought to be raised in principles appropriate to it? If the timidity, chasteness and modesty, which are proper to them are social inventions, it is in society's interest that women acquire these qualities....."

Freud offers remarkably similar justification for women's confinement to domesticity. He writes, "for women the level of what is ethically normal is different from what it is in men. Their superego is never so inexorable, so impersonal, so independent of its emotional origins as we require it to be in men... They show less sense of justice than men, they are more often influenced in their judgements by feeling of affection of hostility..." Hence, Freud insists that the difference in moral capacity between the two sexes must be accepted.

Ironically, most contemporary political theorists continue the same neglect of gender by ignoring the family. Susan Moller Okin has rightly claimed, "the judgement that the family is 'non-political' is implicit in the very fact that it is not discussed in most works of political theory today". A number of examples can be cited.

Recent Development and Approaches

Contemporary feminist scholars have used the word 'gender' after two decades of intensive thought and research. To them gender is a social and political construct, related to and not determined by, biological sex difference.

In its most recent usage, gender seems to have first appeared among American feminists, who wanted to reject biological determinism implicated in the use of such terms as 'sex' or 'sexual difference'. Two major theories of gender are prevalent today-the psychologically focused theory of gender; and the historically and anthropologically focused explanation of gender.

Simone de Beauvoir, in her work *The Second Sex*, a quintessential example of modern feminist enquiry and critique, claims that "one is not born but rather becomes a woman". She claims that It is a whole process by which femininity is manufactured in society. To quote her, "she is defined and differentiated with reference to men and not he with reference to her; she is the incidental, the inessential as opposed to the essential. He is the subject, he is the Absolute – she is the Other".

Beauvoir argues that it is the child-bearing role of women which excluded them from the productive process, and prevented them from seeing themselves as subjects in their own right. Thus, an artificial idea of womanhood was created by society. She exhorts: "No biological, psychological or economic fate determines the figure that the human female presents in society; it is civilization as a whole that produces this creation, intermediate between male and eunuch, who is described female."

Nancy Chodorow also subscribes to the above viewpoint and substantiates her argument from a psychoanalytical point of view. She argues that since women have always been assigned the responsibility of primary parenting and nurturing, they develop a psychology of being more suited to the task of nurturing and caring. Thus she chooses the role of a care-giver and confinement to the private. On the other hand, man from the very beginning is encouraged for more and more individualisation and attaining status leading to personality traits impelling him to associate himself with the pubic.

Chodorow's use of the notion of gender identity presupposes three major premises: First, everyone has a deep sense of self, which is constituted in early childhood through one's interaction with his/her primary parent, and which remains relatively constant thereafter. The second premise is that this deep self differs significantly for men and women, but is roughly similar among women and among men both across cultures and within cultures across lines of class, race and ethnicity The third premise is that, this deep self colours everything one does.

The second theory of gender proclaims that in most societies across cultures, so far, gender has been a socially constructed category rather than biologically determined. However the proponents of this theory have also stressed that the nature of this social construction differs from one society to the other. So, there cannot be any unicausal, universalist and a historical explanation of it. Anthropologist Michelle A Rosaldo supports this theory on the basis of her cross-cultural research, which reveals that women are subjected to the authority due to the existing dichotomy between the public and the private.

Historian Linda J. Nicholson also rejects the unicausal explanation of inequality of the sexes. She emphasises that it has been affected by various causal factors in different social contexts. She stresses the need to fight against the powerful tendency present in political theory to rectify the public-private distinction and to perceive it as rigid. She believes that one can comprehend this distinctly only through the study of history, for the gender structure of a particular time and place is less affected by other contemporary structures, such as political, economic, etc., more by the previous history of gender. Joan W. Scott also stresses the centrality of history in analysing different aspects of the social construction of gender. She explains: a) how cultural myths and symbols reify the suppression of women, b) how these symbols create the 'binary opposition of masculine and feminine male and female, c) how social institutions like family, labour markets, educational institutions and polity reinforce this dichotomy, and d) how the subjective identity formation of individuals is psychologically determined. She, therefore, emphasises the need to expose the social construction of gender by deconstructing it. This calls for:

> "a refusal of the fixed and permanent quality of binary opposition, a genuine historicisation and deconstruction of the terms of sexual difference... (we must) reversal and displace its hierarchical construction, rather than accepting it as real or self-evident or in the nature of things."

In fact, all of them are unequivocal in proclaiming that the public sphere till today, has been constructed under the assumption of male superiority and dominance. It has avoided a number of compelling questions such as that of incorporating the responsibilities of child-bearing and child rearing into the fabric of job-structure.

The above analysis of gender has helped feminists in rejecting many of the existing dominant paradigms. First, they reject the claim that separation of the public and the private follows inevitably from the natural characteristics of the sexes. They argue that a proper understanding of social life is possible only when it is accepted that the two spheres, the private and the public, are inextricably interrelated. Unless the separation of the two worlds is not destroyed, the public life would always be conceptualised as the sphere of men. Carole Pateman has aptly remarked:

> "In popular (and academic) consciousness the duality of female and male often serves to encapsulate or represent the series (or circle) of liberal separations and oppositions: female, or-nature, personal, emotional love, private intuition, morality, ascription, particular, subjection; male, or-culture, political, reason, justice, public, philosophy, power, achievement, universal, freedom."

The most fundamental and general of these opinions associate women with nature and men with culture. Nature is always seen in a lower order than culture. The feminists like Ortner argue that the opposition between women/nature and men/culture is itself a cultural construct and does not exist in nature. S. Firestone in her work *Dialectic of Sex,* argues against this separation of private and public. Women necessarily suffer from a fundamentally oppressive biological condition. It is their role as reproducers that has handicapped women over the centuries and made possible men's patriarchal power.

The consequences of this public-private dichotomy are disastrous. Due to this, women have been deprived of political power and effective participation. Citizenship for women is always seen as 'an elaboration of their private domestic tasks'. Thinkers like Ruskin could argue that "man's duty as a member of the commonwealth, is to assist the maintenance, in the advance and in the defence of the state. The women's duty, as a member of the commonwealth, is to assist in the ordering, in the comforting, and in the beautiful adornment of the state."

This has eventually led to the dichotomy between morality and poor. Women, in the name of being more moral, have been excluded from the public realm. Even the suffragists argued in favour of women's franchise by claiming women's superior morality as it would usher the state in a reign of peace. This is why J. B. Elshtain alleges that suffragists instead of challenging the separation of the public and private, merely "perpetuated the very mystifications and unexamined presumptions, which served to rig the system against them."

Against this background, one can understand the significance of the feminist slogan:

> The feminists argue that public realm of politics can be so rational, noble and universal only because like the messy content of the human body, meeting its needs for production, caretaking and attending to birth and death, as too in politics are taken care of elsewhere.

So, a modern reflective political theory should recognise that the glory of the public is dialectically entwined with exploitation and repression of the private and the people restricted to that sphere so that they can take care of the people's needs. In view of this, the feminist political theory concludes that the 20th century politics requires a basic rethinking of this distinction and its meaning for politics.

The feminists are trying to develop a theory of social practice on the following premises. First, there should be no sexual division of labour at work place and in political organisations of all ideological persuasions; second, there should be a differentiated social order within which various dimensions are distinct but not separate or opposed, and which rests on social conception of individuality, which includes both women and men as biologically differentiated but not unequal creatures.

Third, there is a need to base and expand the conception of politics on the understanding that power relations between men and women are not confined to the 'public' world of law, the state and economics, but pervade all areas of life. This means that contrary to the assumptions of traditional political theory, the family, reproduction and sexuality must be included in political analysis.

The question arises how far this public/private dichotomy and its deconstruction is relevant to the women of the Third World? Can this empowerment epistemology rooted in white women's experiences, subjectivity and identity, fulfil the interest of the Third World women?

However, in the post-colonial discourse, women of the North have understood the significance of the Third World women's realities, and they themselves have rejected the monolithic nature of the earlier feminist discourse on the grounds that it ignores difference, indigenous knowledge, and local expertise. For example, modernity is equated with westernisation, industrialisation and superiority, whereas non-modernity is equated with non-western countries, tradition and inferiority. In the 1970s, Eshter Boserup's pioneering work *Women's Role in Economic Development*, asserted that modernisation had marginalised women and their contributions in the Third World. Chandra Mohanty also questions the western approach, when she asks:

> "Is it possible to refer to the sexual division of labour when the content of this division changes radically from one environment to the next, from one historical juncture to another? At its most abstract level ... concepts such as the sexual division of labour can be useful only if they are generated through local, contextual analyses. If such concepts are assumed to be universally applicable, the resultant homogenisation of class, race, religion, and daily material practices of women in the Third World can create a false sense of the commonality of oppressions, interests, and struggles between and among women globally.

G Sen, and Crown too, are cautious against adopting a concept of gender that ignores differences and diversities. They believe that this diversity is built on gender oppression and hierarchy. He exclaims:

> "Feminism cannot be monolithic in its issues, goals and strategies, since it constitutes the political expression of the concerns of women from different regions, classes, nationalities, and ethnic backgrounds. There is and must be a diversity of feminism, responsive to the different needs and concerns of different women, and defined by them for themselves."

The most positive part of the whole discourse is that these concerns are reflected in practice also. Women involved in different social movements are fighting against injustice on their own cultural terms. Environmental activists like Vandana Shival are speaking of the need to decolonise northern assumptions such as the concept of sustainable development, and appeal to save environment through knowledge based on poor women's experience.

Thus, feminist scholarship has brought sweeping changes to social and political theory. It has cautioned us against constituting gender as a superordinate category of analysis. It should neither be an automatic starting point of analysis of adjustment, nor should it be ignored as a potential starting point. We must avoid approaches that are gender-blinded or gender-blind.

Related Issues

The planners agree that even with expanded employment opportunities, the poor will not be able, with their level of earnings, to buy for themselves all the essential goods and services which should figure in any reasonable concept of a minimum standard of living. The measures for providing larger employment and incomes to the poorer sections will, therefore, have to be supplemented up to at least certain minimum standard, by social consumption and investment in the form of education, health, nutrition, drinking water, housing, communications and electricity, and social welfare services. Social welfare services are intended to cater for the special needs of persons and groups, who by reason of some handicap-social, economic, physical or mental-are unable to avail of or are traditionally denied the amenities and services provided by the community. Women are handicapped by social customs and social values and therefore social welfare services have and should specially endeavour to rehabilitate them by inducing a change in the attitudes of society towards women, their role and contribution.

Various Problems

A statement of a plan of social welfare programmes relating to women, even if, it is within the purview of the overall social welfare programmes, will help in providing the correct emphasis on the problems and development needs of the weaker sections of women and provide voluntary organisations and voluntary effort "a certain" direction. The problems and consequently the developmental action required are, it appears, unlimited and the resources are limited. As such, priorities have necessarily to be assigned.

Among women, the following categories and some of the problems faced by them, call for special attention on a priority basis. The categories are:

1. Working women. To include
   a. The low-income women living in tribal and backward rural areas and urban slums.
   b. The migrant women.
   c. The divorced /separated.
2. Physically and mentally handicapped women.
3. Widows with or without children.
4. Destitute women.
5. Women who come into conflict with law.
6. Exploited women and unmarried mothers.

The problems faced by each of the above categories are numerous and some of them are common to other categories. To decide on action plan priorities, the handicaps and/or the factors which impose constraints need to be understood.

Employed Women

According to 1971 Census, women workers constitute nearly 12 per cent of the total women population and well over 90 per cent of the women workers are found employed in rural areas. It should be recognised here that the problems faced by women workers in rural areas are altogether different from those in the urban areas.

Rural areas including tribal and backward areas: Women workers in rural areas are largely landless agricultural labourers; members of households with uneconomic holdings; those engaged in traditional household industries like hand-spinning, hand-weaving, oil pressing, rice pounding, leather, tobacco processing, etc. These household industries-which are predominantly female labour intensive and which have been a major source of employment in villages appear to have declined in importance during the post-independence period. This is also evidenced by the distinctly declining trend in employment of women workers in the rural areas between the decennial Censuses 1961 and 1971. It has not been possible to reverse this trend because:

1. Almost all the women workers in the rural areas are handicapped by illiteracy and lack of mobility.
2. In addition to this, incessant child bearing coupled with hard domestic work does not provide them any time to go through formal education/ training to acquire new skills. Facilities for acquiring new skills are still sparse.

Urban Areas: Women workers in the urban areas fall into three distinct categories:

1. The first category consists largely of migrants from villages and members of families whose economic position has deteriorated to near starvation. The women of this class work mainly as domestic servants and as unskilled labour in various unorganised industries.

   This category of women workers, who are largely slum dwellers, are below subsistence level. Their problems are to find a job which is secure or provides them regular income at least to subsist, a place for comfortable living, as most of them are away from their homes located in villages; and rehabilitation facilities for their families, particularly children and preparing them for better livelihood, through better education and training; in that order.
2. The second category consists of women, who need employment either to keep their families away from starvation

or to ensure better standard of living. Most of such women are found employed in industries, services and professions. Some are even self-employed. In the case of this second category of women workers in urban areas, their existence ranges from subsistence to security. Some of these women, particularly those residing away from their families, are likely to be exposed to the dangers of exploitation from undesirable and anti-social elements. Personal security is therefore a major problem for them.

3. The third category consists of women who are highly educated and work in higher ranks of services and professions for personal satisfaction and independence. Belonging as they do at least to the upper middle-class families, they do not as group face any serious problems requiring immediate attention here.

   Divorced/separated women are part of each of these categories.

The Physically Handicapped

There are several types of physical handicaps like blindness, deafness, orthopaedic handicap, leprosy, mental retardation, etc., which hinder two persons from even entertaining the hopes of equal participation in the overall social activity. These problems are common to both men and women.

Estimates of physically handicapped women are not separately available. To provide a basis for the formulation of Fifth Five-Year Plan, the working group on the Handicapped constituted for the purpose, estimated that "India may have well over 12 million blind, deaf and orthopaedically handicapped persons. In addition, an estimated 2 million suffer from moderate to severe retardation. The number of persons suffering from leprosy is believed to be around 2.5 million".

The basic problem concerning these physically handicapped persons is lack of adequate facilities for differential medical care, education, training and rehabilitation programmes and a lack of knowledge about these facilities by handicapped persons. Further, it is widely known that though the existing facilities are largely used by men, a majority of physically handicapped women are not coming forward to utilise the available facilities.

The Bizarres

The 1971 Census distribution of women according to marital status indicates that roughly about 9 per cent of the women are widows. Further, they are almost evenly distributed between the rural and urban areas.

Widowhood is a curse for most of the women in India for various reasons:

1. It is almost invariably accompanied by economic disaster. This is because a large number of the families in India survive at below subsistence level and hence death of a male earning member pushes down the families concerned to near starvation. Also many of the females are voluntarily out of workforce, and illiteracy remains the greatest barrier for the improvement of the economic position of widowed women, particularly in the rural areas.
2. Age-old traditions, social prejudices and cultural practices almost exclude widowed women from any socially productive work. Social acceptance of women is reduced with widowhood. In some communities/ regions. There is almost a sort of social boycott of widowed women.

Problems faced by Widows of Different Age Groups: The problems faced by widowed women are not all the same as between different age-groups; and as between rural and urban areas:

1. For widows in the younger age group - particularly those belonging to 15-44 years of age - the problems are more-related to economic independence and rehabilitation in the society - preferably through remarriage.
2. For widows in the age group 45 and above, the problem is more of social acceptance and security. Most of such women, if not previously employed, will be unfit for employment. Even in respect of employed women -widowed after 44 years of age — it is difficult to impart — of the needed training/ skills for more remunerative jobs within the existing framework of education/ training facilities.

Problems of widowed women in the rural areas are even move severe than those in the urban areas. In addition to economic dependence and the social stigma attached to widowhood, there are no opportunities for their emancipation. Many of them are possibly not even aware of the efforts being made by the government agencies through voluntary organisations to redress their miseries.

These destitute women can be classified into three age-groups as their problems are different:

Below 15 years: Persons in this group can be categorised as children. They are mostly orphans and are, therefore, deprived of the tender parental care. They are also subjected to malnutrition and the consequent diseases. This age group, viz., below 15 years of age constitutes the formative years in a person's life, as the process of development and learning are most rapid during these young years. During these formative years, the effects of environment greatly influence the personality development, mental attitudes, moral character, etc. Often, destitute persons in this age group fall a prey to the environmental disadvantages.

15-44 years: This second group of women are both in the productive and reproductive age group. Their main problems are those pertaining to economic independence, social acceptance and security.

45 years and above: In the case of third group of destitute women, their major problem is social security. They are mostly unfit to be employed. They cannot even be trained to earn their livelihood.

Women who fall under this category are: i) juvenile delinquents, ii) women in moral and social danger – particularly those who indulge in immoral traffic and iii) women prisoners.

Juvenile delinquents: are again a creation of the society and the environment in which they are brought up; the deprivation of proper nutrition and training/education which would enable them to earn a better livelihood, etc.

Prostitutes: Women subjected to severe economic distress and hardships often come into the clutches of persons who have vested interests in immoral traffic. Once they succumb, they do not receive proper health care - curative and preventive treatment for the diseases associated with immoral traffic; many of them are not aware of the existing health care facilities and added to it is the innate fear of being exposed to the general public and the resulting social reactions. The existing health facilities are also not adequate and are not perceived as being sympathetic towards their health problems.

Female Prisoners: Many of the problems faced by female prisoners are in common with male prisoners. However, some of the problems are peculiar to females alone. For example, women prisoners with children - particularly in case children are below five years – have problems in arranging for the care of their children. Also, problems in getting rehabilitated, after they are free, are more severe in the case of women prisoners than men prisoners.

Sufferer Children

1. Out-of-wedlock pregnancies are on the increase, judging from the number of abortions and live births among single women recorded at various institutions. Estimates of medical termination of pregnancies (MTP) in the case of single women alone range from 10 to 30 per cent of total MTP cases. In respect of live illegitimate births estimates based on hospital records range from 2 to 3 per cent of the total confinements. In reality many more clandestine live illegitimate births may be taking place which are not brought to public notice.
2. Premarital pregnancies, are as generally believed, no longer confined to the illiterate and depressed classes. According to some case-studies in this field, nearly 50 per cent of the pre-martial pregnancies were observed in the case of women who are at least matriculates. A few were graduates. Some of them were observed to be belonging to the privileged classes of the society. A more distressing feature, however, is that pre-martial pregnancies are being observed even in the case of school girls.

Among the reasons attributed to pre-marital pregnancies are: interactions between various social, psychological and economic forces like bręak down of joint families; overwhelming poverty, rapid urbanisation bringing in its wake the social transformation which leads to increasing permissiveness, lack of communication between children and parents; emotional immaturity and craze for excitement among the youth; antipathy towards the introduction of basic sex education among school children, etc., are the most important reasons cited.

Permissiveness and promiscuity increase with rapid urbanisation and measures to avoid such pre-marital pregnancies is a long drawn social education problem and cannot be expected to decline rapidly. However, the problems concomitant to pre-marital pregnancies can and should at least be tackled effectively.

The action plans should be directed primarily to solving the problems of these six target groups of women.

Practical Standards

There is considerable overlap both in terms of the causes and programmes and agencies concerned with eliminating the problems and building rehabilitation/development plans for these target groups. As such, the action plans are classified under broad groups of actions rather than in terms of categories of women discussed above:

1. Provision of Services/Infrastructure.
2. Education/Training Programmes for the Target Groups.
3. Promoting Voluntary Effort: The Role of Women.
4. Development of Human Resources.
5. Administrative Set-up and Coordination.
6. Legislative Measures.
7. Areas of Research.

The Fifth Five-Year Plan has rightly emphasised the need for a shift in the approach towards social welfare, from a mere provision of curative and rehabilitative services - the kind of approach adopted during the past two decades of planning - to promoting the needed preventive and developmental aspects of social welfare. The action plans should necessarily have such a preventive and development orientation.

Future Strategies

Provision of Services/Infrastructure

1. Services for the care of girls below 15;
2. Facilities for women in the productive age group, i.e. 15-45 years;
3. Programmes for the care of aged and infirm women; and
4. General welfare programmes.

Services for the Care of Girls below 15: There are three categories of children who need particular attention, viz., children of working women, destitute children particularly female children and juvenile delinquents. The following action plans are suggested:

1. The child population below 6 years of working mothers in urban and rural areas is estimated to be around 20 lakh and 166 lakh, respectively. With a view to helping the working mother discharge her duties - both as a mother and worker better family aid services like Anganwadis, Balwadis, creches and day care centres might be launched in a big way.

   Both in the rural and in the urban areas efforts should be made to cover more than 40 per cent of the children of working mothers.
2. There are about 11 lakh destitute children in the country. The girls among them need particular attention, because they are likely to be exposed to social and moral dangers when they grow up. Efforts should be made to provide institutional facilities whether through the foster care programmes or otherwise for taking care of a majority of the destitute female children.
3. It is impossible to discriminate between male and female juvenile delinquents, as the problems are common to both. However, the approach towards juvenile delinquency as such should be to provide the needed atmosphere for a child to develop personality, character and social conscience through setting up of clubs, play centres, juvenile guidance units, workshops, etc.
4. Holiday homes schemes initiated earlier to provide organised and guided recreational facilities to children and be one of

the measures to prevent juvenile delinquency. Such facilities should at least be extended to cover all the children residing in the slums of major cities.

Facilities for Women in the Productive Age Group: In some selected urban areas, hostel facilities are available for working women of the lower income groups earning Rs 50 to Rs 800 per month. However, the coverage of the programme in terms of the proportion of working women needs to be stepped up considerably. Similarly, district-wise investigation would be undertaken about the need for working women's hostels and appropriate facilities set up. The matching contribution for grants for construction/ addition/ alterations should be stepped up.

Socio-economic programmes were initiated in 1958 with the objective of providing full or part-time work to the needy/destitute women and the physically handicapped either through full wage or a wage sufficient to supplement the meagre income of their families. These programmes should be expanded considerably in both rural and urban areas, as they have the potential to provide the needed economic independence to women belonging to the weaker sections and thus act as a preventive measure to many of the social evils.

For effectively implementing these socio-economic programmes, active collaboration should be sought from agencies like Handicrafts Board, Handloom Board, Khadi and Village Industries Board, Small Industries Service Institutes, Small Industries Development Corporations at the State level and the nationalised banks.

To increase the employment potential for the following types of schemes additional steps should be undertaken within the purview of socio-economic programmes:

1. Small-Scale industries.
2. Units as ancillary to large production of handicrafts.
3. Units for the procurement and production of handicrafts.
4. Handloom training-cum-production units.
5. Agro-based industries like dairy, poultry farms, etc.

6. Traditional female labour intensive industries like rice pounding, oil-pressing, etc.

Attempts must be made to revitalise and activate the existing sick units falling under the purview of socio-economic programmes.

Ways to improve the working, efficiency and effectiveness of Mahila Mandals must be studied and necessary action taken. They should be reoriented to aim at increasing the earning power of women in the rural areas.

It is suggested that by the end of the Fifth Plan, about 10,000 Mahila Mandals, should be developed throughout the country to provide an effective media for organising women welfare activities in the rural areas.

Scholarship programmes of the Central and State Governments for the handicapped should be expanded considerably and efforts should be made to encourage women to make use of the scholarships available.

Sheltered workshops should be organised.

Schemes for the welfare of destitute women between the ages 18-44 and 45-65 providing for basic amenities of food, shelter, clothing, basic education and training in crafts should be implemented through voluntary organisations who may be given grants to cover 75 per cent of the expenditure. It is suggested that this scheme should be revived and implemented in all the States.

Homes for the rehabilitation of rescued and released women prisoners should be started in all towns with a minimum of 5 lakh of population. Apart from providing shelter, food and clothing, the inmates should also be provided training in crafts like sewing, embroidery, knitting, etc. Efforts should, however, be made in the direction of making inmates self-sufficient and earn independent livelihood.

In some such protective homes, insane women are housed along with other women which is an unhealthy and undesirable practice and should be discontinued.

Programmes for the Care of Aged and Infirm Women: Women in the age group 65 years and over constitute roughly 85 lakh according to the 1971 Census. Many of the women lack absolutely any security They are mostly dependent on their children who often desert them. Thus, even women belonging to upper middle classes are sometimes reduced to the status of destitute. Efforts should therefore be made at least in a modest way to initiate social security measures through old age pension with the objective of providing economic independence to at least 25 per cent of women in the age group particularly in the rural areas.

For the women retired from active service and for those who are in need of some residential facilities, hostels should be started in all the major cities. If necessary, subsidies may also be extended under the grants-in-aid programmes.

General Welfare Programmes: Slum clearance programmes should be initiated in all the major cities and towns with a minimum of 5 lakh of population. People displaced should be provided alternative sites, with proper environmental sanitation, for building their homes.

Zila Parishads and youth in the districts should be entrusted with drinking water supply projects.

A vigorous campaign of education and action should be launched in favour of community sanitation and hygiene. Public utility services should be expanded. The practice of carrying night soil as headloads must be eradicated.

Education/Training Programmes for the Target Groups: Analysis of the problems faced by the target group of women indicated that illiteracy, inadequate education/training, lack of facilities for training in alternative skills and lack of knowledge about the existing facilities are some of the major problems that have hindered the progress of women in India. There is, therefore, the need for accelerating the efforts in this regard with renewed vigour. With this in view, the following action plans are recommended:

(1a) The Fourth Plan introduced a programme of functional literacy built round farmer's training in selected districts where high yielding varieties of crops were being cultivated. It is estimated that about 90,000 women received this training during the Fourth Plan and about 5 to 7 lakh of women are likely to be trained under this programmes during the Fifth

Plan. This programme must be extended to all the rural areas.

(1b) Apart from imparting knowledge about farming, the curriculum for women should include courses of training, in occupational skills like kitchen gardening, food cultivation, poultry keeping, animal husbandry; household arts like cooking, nutritional values of foods locally available, sewing, knitting, etc.; and family planning.

(1c) Preference should be given to women belonging to Scheduled Castes, tribal women, widowed women and destitutes under this functional literacy programme.

(2a) For the non-student young girls without any education and school drop-outs — particularly for girls in the age group 11-14 years, the pre-vocational training programmes should be reviewed and strengthened by enlarging the scope of training and by increasing the number of trades.

(2b) In respect of girls in the age group 11-14 years, the objective of pre-vocational training should also be to train them to be self-sufficient in home management by organising courses of training in sewing, cooking, nutrition, minor repairs of the house, motherhood, child care, etc.

(2c) Pre-vocational programmes should be extended to cover girls in this age group in the rural areas. In urban areas, preference should be given to girls in the slum areas and destitute girls.

(3) Condensed courses of education were started in 1958 with the twin objective of: a) opening new vistas of employment to a large number of deserving and needy women, and b) creating a band of competent trained workers required to man the various projects in the rural areas in the shortest possible time. Under the scheme, women in the age group 18-30 who have studied up to classes IV and VI are trained for middle school/matriculation examinations within a period of two years. The scheme was found very useful but the statistics reveal that the beneficiaries have been mostly women belonging to the middle-class families. Preference should be given to women belonging to backward classes, widowed women and destitute women.

(4) Special efforts should be made to cover women belonging to Scheduled Castes and Scheduled Tribes through condensed courses. An incentive of Rs 1,000 (as recommended by the Review Committee), be given to the institution for every successful Scheduled Caste/Scheduled Tribe candidate trained.

(5a) The condensed courses should be organised in a big way and for smaller groups of say 5 to 7 with the help of high schools and colleges for girls. Efforts should be made to cover about 215 lakh women under the condensed course programmes, during the Fifth Plan period.

(5b) Apart from imparting general education, condensed courses should also aim at imparting job-oriented training with the active cooperation of existing vocational training institutions.

(5c) Under this programme of condensed courses, short-term courses should be organised to retain women who have been temporarily out of job-market to fulfil child bearing responsibilities.

(5d) For the failed candidates, short-term course of six months to one year should be organised.

(5e) Special efforts should be initiated to follow-up successful candidates with a view to helping them in securing jobs.

(6) Pre-examination training facilities should be offered to duly qualified poor women with the objective of equipping them to successfully compete in examinations for public jobs. It is suggested that about 80 lakh girls in the age group 14-17 may be covered under this programme during the Fifth Plan period.

(7) The school curricula in various States in India should encourage the doing away of traditional prejudices of inequality of the sexes.

(8) Sex education should be introduced at the appropriate stage with the objective of also educating the young girls about the social and moral dangers they are likely to encounter.

(9) The value of physical training in the school curricula should be emphasised.

Promoting Voluntary Effort: The Role of Women: Voluntary welfare service organisations have been an integral part of the cultural and social traditions in India. Soon after independence, it was estimated that there were 10,000 voluntary organisations engaged in social welfare. In fact, all the schemes of Central Social Welfare Board are implemented only through voluntary organisations. The reorientation given to social welfare in the Fifth Plan calls for more effort on the part of both voluntary organisations and the State agencies involved. The following action plans are, therefore, warranted in this regard:

1. Efforts should be made to promote a large number of voluntary organisations throughout the country. They have a critical role in mobilising public opinion in favour of equality among men and women, and eradicating superstitions, social evils and waste. The motivational strategy for encouraging voluntary organisations needs to be well thought through and support facilities provided. Women should be promoted to take the initiative and responsibility for organising voluntary effort, for not only can they bring to the tasks the necessary dedication commitment and empathy; but their very presence will provide their socially handicapped sisters a source of inspiration and set in a cycle of social rejuvenation. All voluntary organisations particularly those concerned with social welfare vis-a-vis women must be encouraged to have women members. Women Panchayats, Mahila Mandals, working women, etc., should be encouraged to spearhead such voluntary activities. Mahila Mandals should be promoted in every village so that they can function as field level agencies for social and economic transformation.
2. Most of the voluntary organisations have been operating independently of each other. They have, therefore, not been able to fully benefit the community. The role of existing organisations should be determined and measures should be initiated to coordinate/supplement the efforts of various organisations at each district level.
3. Many of the women's voluntary organisations are located in urban areas, while only a few organisations have endeavoured to work amongst rural women. Efforts should be made to

promote a large number of voluntary women's organisations in the rural backward and tribal areas and urban slums to mobilise public support for different programmes and to implement them. This calls for liberation of the rules regarding the matching grant through voluntary contributions, simplification of the rules and procedures of obtaining the grant as well as administering the organisations, provision of trained staff, organisation of leadership training programmes, etc.

Development of Human Resources: Administration of various social welfare programmes have become increasingly technical. During the past two decades of developmental planning, lack of technically competent workers has had an adverse impact on the quality and success of welfare programmes. With a view to provide the necessary support to various agencies, the following action plans are suggested:

1. Training facilities for the workers attached to all the voluntary agencies, like Mahila Mandals should be initiated immediately. The training needs of workers, however, differ from organisation to organisation depending on the nature of tasks required to be performed.
2. Through a proper investigation training requirements of workers in each district should be assessed and suitable training programmes designed.
3. These training programmes should, as far as possible be organised at each district level.
4. Trainees should preferably be local candidates.
5. Effective implementation of the various socio-economic programmes require two cadres of workers: the grass-root workers and supervisory staff. The grass-root workers should be provided training in the latest techniques and methods of production with the active collaboration of well established industrial units and Industrial Training Institutes. The supervisory staff, on the other hand, should be trained in advanced techniques of production, business management, personnel management, etc.

6. In the case of handicrafts units under the socio-economic programme, practising craftsmen should be trained as instructors and appointed.
7. Short-term orientation should also be given to the members of the managing committees of the units – socio-economic programmes - about the general working of such units.
8. Senior level officers in charge of the socio-economic programmes should also be exposed to short-term orientation courses in business management and allied fields through Small Industries Service Institutes, University departments of business management, etc.

Administrative Set-Up and Coordination: Administrative traditions in India have tended to attach least importance to departments dealing with social welfare. This is reflected even in the training imparted to administrators. Only recently it has been realised that administration must also be welfare oriented. The federal nature of our policy vests a large responsibility for implementing social policy and programmes with State and local authorities. There is, therefore, the need for reorganising the administrative set-up with a view to effectively implementing the various welfare programmes. The following action plans may be taken up for consideration:

1. Orientation/training programmes should be organised for social welfare personnel, particularly at decision making levels, to sensitise them to social welfare needs and adopt the extension approach of reaching out to the clients. The new developmental and preventive concept, of welfare also needs to be imparted.
2. Every State Department of Social Welfare should have a Women's Welfare Division with responsibility for planning, programming and monitoring the implementation of schemes of women welfare.
3. The Central Social Welfare Board is one of the most important agencies for the implementation of social welfare activities. It should be reorganised and strengthened, and vested with larger funds and responsibilities for promoting and developing voluntary effort particularly in rural, backward

and tribal areas and among the weaker sections of the community.

4. The Central Social Welfare Board should launch a massive campaign for enlisting and developing a cadre of voluntary social workers who should be provided some normal assistance to enable them to carry out this work.
5. State Social Welfare (Advisory) Boards should also be reorganised and strengthened.
6. The State Board should also be made to function as liaison among the State Government and the local agencies.
7. Suitable infrastructure should be developed at each district level and block level for implementing and expanding the programmes of Central Social Welfare Board.
8. Trained social welfare workers should be associated with all the committees to be set-up by the Central Social Welfare Board.

Legislative Measures

1. International experience indicates that evolving a sound social security system takes a long period of time. However, suitable enactment can be initiated to provide public assistance to select groups like destitute women and people above 65 years but without any means of livelihood. Assistance here need not be in the form of cash. It should be in the form of medical, housing, feeding and recreational facilities, etc.
2. It should be open to the States and Union Territories to go in for taxation or special levy to finance such public assistance schemes without prejudice to any assistance made available to States and Union Territories from the Central Government under plan schemes.
3. No child should be tried in adult courts nor should any child be sent to a jail.
4. State Governments should enact legislation for apprehension, institutional treatment and rehabilitation of beggars, particularly women.
5. Machinery should be set up for speedy and effective adjudication in all cases concerning the family, including the

setting up of family welfare courts since the ordinary judicial procedure is not suited to handle such cases. Women, particularly in rural areas, should be protected against harassment.

6. A vigorous campaign should be launched to educate women about their rights and the machinery through which they can seek their realisation.
7. Active public support should be mobilised by government agencies, voluntary organisations and public leaders against child marriage and dowry to support the legislative measures for the eradication of these undesirable practices. Ostentatious weddings and other wasteful social ceremonies should be banned.

Fundamental Variations

The term societal resilience is to be understood in its contextual sense to be meaningful as a framework for analysis of social processes in the developing societies. The context lies in the past history of colonial exploitation and neo-colonial threats to these societies, and the continual attempts by imperial nations to dominate their economic, political and cultural autonomy. The fragility of the institutional foci in these nations gives a poignant meaning to the notion of resilience.

The, resilience of societal structure and tradition connotes the ability of the developing nations to withstand socially, culturally and economically the forces of instability both internal and external to their system as they plan their march towards modernisation. Obviously, the term societal resilience is to be understood not in a literal sense; such as status quo or static contextulity in the system. It is to be used as the contextually relevant notion or even a process through which the crises of cultural, economic and political identity of the 'new' nations in Asia, Africa and Latin America are articulated leading to the evolution of a meaningful strategy for their autonomous process of development and modernisation.

The historicity of social transformation in each society in this context assumes great significance. The process of modernisation needs to be analysed in the background of history and tradition of each society to fathom the potential of its societal resilience. This is important because it is observed that the initial impact of the forces of modernisation generates social and cultural tensions in most societies which put to test their capacity for societal resilience. Otherwise, it leads to cultural and institutional break-down in societies resulting into their dependency upon 'big powers' leading to neo-colonial relationship of subordination. Consequently the democratic processes are subverted, cultural institutions are debased and a psychology of disenchantment from one's own tradition comes into existence and becomes a source of ever widening cultural alienation. The nations that get caught into this relationship of, neo-colonialism have certain institutional and social historicities which need to be analysed to grasp the significance of societal resilience for successful modernisation.

Social structure and its historical specificities influence the degree of resilience that a social system has in withstanding the challenges of development and modernisation. In the Indian experience, such historical forces were defined by inter-structural autonomy within the society. The impact of initial forces of technological, institutional and cultural modernisation coming through colonialism was encapsulated by the social system because of the autonomy among its major structural foci, such as the polity, social stratification and value system.

The inter structural autonomy in the Indian society has subsisted through its tradition and past history. As a result, the exogenous impacts, political, economic and cultural, were moderated or even encapsulated in the system through selective adaptation which helped their indigenisation instead of their creating forces of disenchantment from the native tradition. The traditional Indian society was characterised by substantial autonomy between the roles of the literate or priest who served as spokesman for dominant values and cultural practices and the political rulers (or Kings) who held power. In a symbolic form, this is, exemplified in the Indian tradition by the autonomy or even superiority of the 'priest' over the 'king' in matters concerning tradition or its interpretation. This autonomy added to the resilience of the Indian society and its tradition. Even if the political institutions came under stress or were overrun the cultural tradition maintained its autonomy of form and functions. Just as the priests did not enjoy legitimacy for influencing the functions of the political institutions (except in rare instances), the ruler too were subjected to compliance to the priests in matters of normative pronouncement on values and tradition.

Similarly, the traditional institution of social stratification, which was characterised in India by the varna-caste hierarchy, functioned almost autonomously through the caste, panchayats, jajmani networks and its self-governing guilds. The caste panchayats dealt with most matters related to the economic, social and cultural practices, customary rights and duties and in rate or exceptional cases did the members of a caste approach the court of the King or the state for settlement of disputes. This gave caste system not only a regional character but also cultural, economic and social autonomy from both the interference of the political rulers as well as the Brahmin priests. Indeed, the relationship between the priest and the castes in traditional India was of curious sociological significance. Priests did not enjoy a uniform access to roles in ritual practices of each caste as such. Their relationship with the lower castes and outcastes was always marked by segregation; they did not enjoy, uniform respectability from a large number of middle-peasant castes. Only in relation to the upper castes did the, priestly calling have a relationship of sustained ritual, interaction. Even here, the priests were accorded superior, status only in matters relating to the ritual practices. In the realm of power and social hierarchy a rich member of the upper caste, say from the ruling clans would not accord the priest a superior status. In such matters, ritual obeisance to the priest did not accord with higher social status of the priest.

These structural features of the traditional Indian society rendered it possible for the society to assimilate new external values and institutions selectively without being swept over by the impact of their cultural, political and economic forces. It enabled the social system to maintain are remarkable degree of continuity with change. This process was put to severe test following the establishment of the British colonial rule in India. The contact with the western cultural tradition through colonial rule was for India in the nature of an encounter of a historical kind. It represented an encounter of a technologically and scientifically stronger tradition inhering modernisation values of rationality, enterprise and industrialism with one which was suffering from its inner tensions and institutional disarray. Such an encounter, in a colonial setting could have swept India from its cultural roots but for the resilience of its social system resulting from the historicity of its social structure and tradition. The impact of westernisation in the initial stages of the colonial rule was moderated in India through a process of structural encapsulation. This is evident from the fact that the educational, economic and cultural policies that the British introduced in India during the 19th century only succeeded in creating a middle class largely confined to the metropolitan cities, big towns and upper castes.

The social background of this middle class, which emerged as a result of the colonial form of modernisation had features which initiated the freedom movement on the one hand, and on the other, encapsulated the impact of the western cultural onslaught and tensions of the colonial penetration. Thus, the social and cultural forces such as values of equality, legal rationality, technology and science which this western contact generated were structurally confined to a limited group and the system gained time for slow percolation of these values to the rest of the society through an adaptive process of transformation. It did not cause major dissociation in the established institutions, values, social structure and tradition.

Eventually, the growth of national independence movement which in the beginning was elite sponsored, gained a mass character under Gandhiji's leadership. It made it possible for an alternate ideology for India's social, economic and political modernisation to evolve with wider sensitisation of people to this ideology. The historical forces in the society imparted a social resilience to the system that it could after independence undertake a massive and revolutionary programme of political, social and economic modernisation despite the human cost of partition. The partition, was the result of the colonial design to subvert Indian forces of nationalism and battle against imperialism. The colonial policy of divide and rule for a moment proved stronger than traditional forces of resilience in the system and the country was partitioned on communal principle of two-nation theory which Indian leadership refused to recognise and even today does not accept.

The second phase of modernisation began with the, achievement of Independence at the end of the colonial rule. Long before this, the Indian national movement had contributed to the growth of modernisation and development ideology for the Indian society. As the national leadership took over the reign of the government,' this ideology was given a paradigmatic shape through the Constitution.

The paradigm of modernisation derived many elements, from the Western tradition, such as the emphasis on rationality, secularism, parliamentary form of democracy, adult franchise and a rational judicial and administrative system, etc.

The Constitution, however, also incorporated many elements from the Indian nationalist movement and history, such as emphasis on social justice, moral bases of the directive principles of the state policy, the provision for protective discrimination for scheduled castes and tribes and special provisions for the religious minorities, backward classes and women. It also laid emphasis on maintaining the positive aspects of the Indian culture and tradition in the process of economic development and modernisation. Thus, the paradigm of modernisation which India adopted after Independence was itself a measure of resilience of its social structure and values. At the same time it set into motion forces of transformation in society which increasingly put the burden of proof about societal resilience on the historicity of its processes and their social framework.

Historical Ideas

The first issue that emerged soon after India embarked on the path of parliamentary democracy with adult suffrage was as to whether its people, illiterate and traditional as they were, would be able to absorb the tensions released in the process of electoral participation. Could parliamentary democracy succeed in a society where most people were unlettered? While posing these questions, it was overlooked how a people with centuries old heritage of civilization also intuitively imbibe skills through enculturation and non-formal education that serve as instruments both of succorance and catalysis in discerning political participation. The erroneous assumption was to equate 'illiteracy' with lack of 'education'. The 'education' to which in a broader sense the people were already exposed to through mass participation in the national movement was overlooked because the analytical model was imitatively borrowed from the West.

The electoral politics and the level of political participation in India on the contrary has shown remarkable resilience through its responsive character and feed-backs from interest groups on issues of devolution of powers, social policies and finer electoral discriminations in respect of voting for Parliament and the Assemblies. Even though the structure of party system in the country has, not yet stabilised, a framework is already emerging of coalitional strategies between the regional and national political parties which is in consonance with the Indian social and cultural pluralism and its historicity. The process is yet in the making.

Yet another negative prognosis on the role of tradition and traditional values in the modernisation of Indian society was based on the argument of the other-worldly orientation of Hindu values which were expected to retard successful industrialisation and economic growth. This too has been proved to be wrong as the evidence from the pattern and direction of entrepreneurial activities and people's active participation in industrial work in all parts of the country suggests. In the villages, even the puritan Brahmins have not hesitated from sericulture (which meant killing of silkworms, a practice not approved by tradition) when they realised its economic advantages.

The industrial workforce, even those producing traditionally polluting commodities such as the leather goods and chemicals, has turned out to be fully representative of the social spectrum comprising all castes and creeds. This falsifies the assumption that caste and religion would hinder the process of industrial modernisation. Studies of urban centres in various parts of the country have established that religious gatherings such as the 'Hari Kathas' and prayer assemblies not only preach religion but also serve as means of communication of messages relevant to modernisation ideology of the country, such as social reforms, upliftment of the status of women, curbing the practice of caste discrimination and other social evils. Such media have also been made use of by contending political groups and parties, even the parties professing radical left ideology. Such uses of the traditional institutions and values in the promotion of modern institutions and values have established a successful strategy of modernisation with the continuity of tradition.

This is further evident when we observe the role that traditional joint family structure has been playing in country's economic and industrial modernisation. Even during the British period when the business houses in India were not encouraged to establish industrial units in competition with the British concerns, and when credit was not available from the government sources, the joint family and kinship resources of the members of the business community were successfully utilised for raising capital for investment in industrial units in the country. This has been specially true for the cotton textile mills in Maharashtra and Gujarat during the early twentieth century. The same pattern has also continued after Independence despite governmental support for capital funding as an organic element of the planning process. The studies of industrial entrepreneurships, their life-cycles, rise and fall as conducted by social anthropologists have revealed that there exists a causal relationship between successful functioning and growth of an industrial unit and the internal cohesiveness among the members of the family which owns and manages the enterprises. There are instances of disintegration of industrial entrepreneurial units following partition in the joint family of the entrepreneur households, It is also found that managing agency' system of ownership of industries or firms in India has its basis in the family structure.

A more prominent relationship between the joint family structure and economic development in India has been highlighted in the studies of the agricultural development called the 'green revolution'. It has ensured not only food self-sufficiency but also food surpluses to the country. Studies of social structure and values system of the farmers involved in green revolution bring out interesting sociological features which establish the case for continuity of tradition in the process of economic modernisation. The leaders of green revolution in all parts of the country come from the traditional peasant castes, normally the middle castes which have had centuries old tradition of cultivation on land as their calling. The mode of their work on the farm holdings invariably is oriented to the joint-family, partnership. This not only ensures efficiency but also economy in the process of production. The joint family mode of Production nevertheless also selectively includes elements of rational uses of technology, such as uses of new variety of seeds, fertilizers, pesticides and rational methods of irrigation essential for higher yield.

It also makes use of modern techniques of marketing, banking and credit along with utilising the resources of traditional family and kinship ties. This is combined with continuity of such traditional values of peasant castes as frugal style of living, consumption mainly of the traditional items of food, clothing and living, etc. and careful family budgeting. These contribute to higher economic surpluses which are used not only for further economic modernisation and expansion, but also for incurring expenses on such traditional items as marriage feasts, construction of houses, temples and sacred ponds, etc. in villages. All this suggests a curious amalgam of tradition with modernity in the modernising economic and social profile of the countryside in India today.

These sociological facts amply reveal the continuity of, tradition in the process of modernisation of economy, social structure and value system of the Indian society. Such adaptive tendencies have been consistently observed in the emerging functions of institutions of religion, caste family and kinship and folk cultural tradition. It is a measure of the resilience of the traditional institutions in India that most of them have been successful in effecting a selective and rational adaptation to modernity, without, creating major social hiatus in its social structure or tradition. This process has prevailed in India throughout the 1950s to 1970s. This period can be identified as carrier of the second phase of the modernisation process in India.

As we mentioned earlier, the first phase of modernisation coincided with British colonialism when this process was largely confined to the urban middle cl[illegible]es and existed in an encapsulated form. The period after Independence enlarged the social base of modernisation and set into motion social, economic and political forces which had far reaching consequences for all segments of the society. The policy of protective discrimination for the scheduled castes and tribes together with adult franchise in their electoral participation energised them for new scale of social participation and social mobilisation.

It also generated social conflict which could largely be accommodated in the social system not only through progressive incorporation of these groups in the political processes of the society but also through implementing the policies of distributive justice. Soon after Independence caste-vote-banks played significant role in influencing the political and electoral behaviour of Parties and people in India. It led to politicalisation of castes, emergence of caste associations and backward caste and class movements. A surface view on these developments might make them appear like articulation of traditional institutions or their re-invigoration. A deeper evaluation of their structure and functions, however, reveals a much different story. These developments represent not the traditionalisation of the forces of modernity in society, but modernisation of its tradition.

The second phase of modernisation can be marked for its focus on capital investment for industrialisation, growth of technology, science, education, agriculture, energy and economic infra- structures, etc. on the one hand, and on the other, major social and economic reforms. The abolition of zamindari, agrestic serfdom, tenurial reforms, ceiling on land holdings, etc. explicate nation's commitment to establishing an equalitarian society. The nation-state with its ideological commitment to secularism, democracy, social justice and pluralistic values reinforced its legitimacy through these reforms. The new investments both in terms of capital resources and innovative institutions accelerated the process of social mobility and also generated social conflicts. Both were inevitable, because society was passing from traditional state of its static equilibrium to the state of dynamic equilibrium through social and -economic modernisation. Consequently, the pressure of interest groups, caste associations and backward classes increased for social mobility and economic development as a large segment of the deprived groups desired such opportunities traditionally. As a result of these processes the social system as a whole witnessed some major changes such as the abolition of landed aristocracy, emergence of middle caste and middle class dominance in urban and rural parts of society because of the opportunities made available to them through development planning initiated by the state, growth of new entrepreneurial classes in cities in sectors of small scale industries, trade and market enterprises and noticeable upward social mobility from among the deprived castes and classes as a result of protective discrimination in their favour.

Even though its incidence was often localised, was uneven or differentiated compared to mobility among other social groups and classed, in qualitative terms it marked a major social change. The extent to which these castes and classes witness improvement in their social and economic condition remains a matter of debate among social scientists.

Study Area

Primary data available with sources such as the Census and National Sample Survey, are insufficient and are very scanty for social welfare planning, particularly on the needs and requirements of handicapped women, destitute women, women under the purview of the suppression of Immoral Traffic Act, etc.: In view of this, the following areas of research are suggested:

1. Studies on 'Social profiles' with district as unit' wherein information on the prevailing conditions of social needs and requirements, etc., are investigated.
2. Studies on the requirements of physically handicapped children and women.
3. Studies on the requirements of destitute children and women.
4. Studies on the training requirements of workers in voluntary welfare organisations.
5. Studies on the socio-economic and psychological factors behind the problem of pre-marital pregnancies.
6. Studies on the magnitude of problems facing prostitutes and their children such as problems of children of prostitutes, particularly female children.

NGO's Role

Voluntary action in India has always been an integral part of the cultural and social traditions. A variety of social services were provided by voluntary agencies prior to independence and in the first few decades' of planned development in India. Traditionally, voluntary agencies undertook a wide variety of activities in the areas of social reform in the pre-independence period. Independence resulted in government policy and commitment to support and strengthen voluntary agencies. Voluntary agencies have currently opted for several alternative roles depending on their objectives, location (rural/urban) and resources available.

The role of voluntary agencies in national development has been considered vital due to their direct and first-hand experience and knowledge of local needs, problems and resources at the grass-roots. Further the commitment and zeal of the voluntary action movement is considered effective as it is not bound by rigid bureaucratic systems and is more responsive to people. The voluntary sector is observed to operate with great flexibility and bases its activities on felt needs. There is a process of continually learning from past experiences in programme planning and implementation, etc.

The essential strength of voluntary agencies derives from the fact that they are closer to the community and people. They represent in many cases the needs and aspirations of the people. Voluntary agencies often function more effectively than the government managed agencies in areas such as motivation, problem identification and analysis, project formulation, innovative methods of service delivery and involvement of the community, due to their spirit.

There are a number of lessons to be learned in such areas such as the demystification of technology; de-emphasising formal educational qualification in favour of experience, capabilities, aptitude and ability to work with people; expansion of activities without adding on cumbersome bureaucracy; and reliance on community based and non-institutional approaches. The unique strength of the voluntary sector is its ability to pressurise the government without succumbing to it and losing its identity and lobbying on issues and ideas to make them acceptable to government and the people. The decentralised administration in the voluntary sector not only facilitates effective grass-roots, delivery mechanisms but also ensures the participation of the beneficiaries in the programmes.

Voluntary agencies in India have evolved as a result of a historical process that has brought them to their present status and role in the country's development. In the 1950s, most of the organisations provided either relief work or were involved in institutionalised programmes such as schools, destitute homes, hospitals as well as welfare activities. In the 1960s, many of these organisations realised that families with a weak economic base would be unable to procure the benefits of institutional welfare and relief services. It led them to the conclusion that services should enable beneficiaries to be productive and self-reliant through income generating programmes. In the 1970s, many of the voluntary organisations began to feel that economic inputs alone could not overcome poverty and a critical roadblock to development was the unequal social structure. A new type of education geared to raising the consciousness of weaker sections on their situation and rights so that they become active agents of their own development, and change was considered essential. Activist groups built around these considerations, subsequently came into existence in the voluntary sector.

The organisation of women by voluntary organisations has acquired importance as the need has emerged for organisational structures to ensure women's participation in the development process. Many old established voluntary agencies have undertaken the task of setting up welfare development services for women in the country.

New Trends: From the mid 1970s onwards there was an emergence of many newly established organisations and activist groups. A large part of the activities of these groups have centred around combating atrocities and violence committed on women, dowry murder, brutal forms of maltreatment and exploitation. In many cases, women in distress have approached such groups for assistance in registering and follow up of cases, providing shelter, etc. These activist groups have identified themselves with oppressed, victimised, and harassed women and awakened new hopes, aspirations and consciousness among women on these issues. Recently many formal and informal groups have emerged throughout the country which have successfully mobilised women's awareness and have preferred to work directly with the women, relying less on material inputs from the outside and more on increasing the internal capabilities and resources-economic, social, cultural and political. These activist groups have also elicited the intervention of the State, especially of the judiciary and of the fourth estate, to project the rights of women and ameliorate their situations. At the same time, they have organised the women themselves for struggle.

Besides voluntary agencies and activist groups, there are many other functional groups such as Mahila Mandals, Youth Clubs, Nehru Yuvak Kendras, National Service Schemes, cooperatives and other people's institutions that have effectively taken up the issues of women in development with varying degrees of success.

Government's Stance on Voluntary Action: The Planning Commission has recognised the role of voluntary action in accelerating the process of social and economic development in most of its plans, particularly so in the Sixth and Seventh Five-Year Plans. Voluntary agencies at their best have played an important role in providing a basis for testing and devising innovative projects and new Models and approaches in programme implementation and in ensuring feedback, as well as in securing the participation of women living below the poverty line. They have developed competence in many non-traditional areas and played a vital role in supplementing governmental efforts so as to offer the rural poor choices and alternatives. They have often served as the eyes and ears of the people at the village level. By adopting simple, innovative, flexible and inexpensive means to suit their limited resources, they have tried to reach a larger number of beneficiaries with minimal overheads and with greater community participation. In the process they have successfully demonstrated how village and indigenous resources, rural skills and local knowledge are grossly underutilised at present, in a cost-effective manner. Voluntary agencies have also managed to mobilise and organise the poor to some extent and to generate in them the awareness to demand quality services and improve accountability of the local level functionaries. They have helped to train a cadre of grass root workers that believe in professionalising voluntarism.

The increasing interest of the government in enhancing the role of voluntary agencies in the development of women is quite evident. Considering the magnitude of problems faced by women the government has rightly felt that it cannot assume the entire responsibility of service provision and development. It has sought to associate voluntary agencies in the various programmes aimed at women. The thrust of the current programmes is more towards development of women's potential and their productive participation in development rather than merely providing welfare services to them. A meaningful partnership with the voluntary sector has thus been an avowed goal and an essential variable in government's attempts to integrate women in development.

Women and the Voluntary Sector: Voluntary agencies have contributed immensely to the new directions and impetus provided to women's programmes during the decade for women. A number of innovative features in several government formulated schemes/ programmes are based on the experience of the projects run successfully by voluntary agencies.

The rationale for involvement of voluntary agencies in women's development is quite clear. Women in India suffer from multifarious constraints such as a low level of literacy, lack of access to resources and obstacles caused by the cultural and social customs and traditions that are discriminatory of women. In a situation such as this, the role of voluntary agencies in creating awareness among women of their rights and mobilising women as well as developing in them appropriate motivation and leadership to realise those rights cannot be minimised.

The process of creating an environment conducive to the progress of women is dependent on a multitude of socio-economic factors, starting with a political will to enforce the development of women as a priority. The long-term objectives of the Seventh Plan spell out that raising the economic and social status of women is a critical goal of national development. The basic approach suggested is to inculcate confidence among women and bring about an awareness of their own potential for development. Within this framework, gainful employment to women is accorded the highest priority as an effective strategy. Various ministries and departments have formulated programmes for the development of women with an emphasis on the involvement of voluntary agencies as delivery mechanisms. The role of voluntary agencies in the mobilisation of women in particular is seen as a critical factor for the development strategies of the future.

A higher involvement of voluntary agencies is thus envisaged in the implementation of such government programmes as the Integrated Rural Development Programme (IRDP), Training of Rural Youth in Self-Employment (TRYSEM), Development of Women and Children in Rural Areas (DWCRA), Integrated Child Development Services Scheme (ICDS), and Adult Literacy programmes. Besides their involvement in these schemes, voluntary agencies can also assist in effective enforcement of minimum wages, supply of safe drinking water, afforestation, social forestry, consumer protection, promotion of science and technology, rural housing, legal education, etc. With the new focus on women, some funds should be earmarked for implementation of these programmes in the concerned ministries/ departments for voluntary agencies. Further through the Central Social Welfare Board (CSWB), Council for Advancement of People's Action and Rural Technology (CAPART) and the National Rural Development Fund, the activities of and cooperation with voluntary agencies should be expanded and strengthened. To the extent that voluntary agencies are dependent on public funds, accountability has to be ensured but without cumbersome and rigid methods.

Voluntary Action in the Organisation of Women: Empowerment of women cannot be ensured until they are enabled to organise themselves. Collective organisations spell strength. This is a prerequisite for initiating action, lobbying, pressurising and bargaining. Grass roots organisations can greatly enhance the opportunities for poor women to participate in development programmes by providing an organisational base to operate from. By organising, working together, sharing experiences and resources, building pressure groups and so forth, women can find independent access to opportunities for their betterment.

A large number of women are engaged in the unorganised sector working and living under precarious conditions and with no legal protection. The unorganised sector denies women all benefits of collective action. Dispersed and unorganised, they have no political power and no bargaining strength. As a result, it becomes much more difficult to implement protective labour laws relating to wages, conditions of work, insurance, provident fund, maternity leave, and creches, etc., and also to channelise economic inputs such as credit, technical training and marketing. In such a situation, the need for collective action becomes critical, and is dependent upon the organisation of women in the unorganised sector. Many spontaneous and organised struggles have been launched by some voluntary organisations for the articulation of the needs of poor women, particularly the need to organise them for their interaction into the mainstream. For instance, the whole issue of women in the unorganised sector has been debated and seriously addressed through the awareness generated by certain organisations in different parts of the country.

Uncovered Territory: The issue of gender disparity at work is yet to be voiced effectively in the organised voluntary action movement. Of the vast masses in the category of the working poor, the unskilled ranks contain a larger proportion of females. These women are much less organised for any kind of market leverage or wage bargaining and even when organised, less inclined to redress gender inequalities at work sites. There is the need to replicate the success stories of voluntary action in organising women in different parts of the country and to take up the issue to parity at work in a larger way.

The participation of women in development requires an all round transformation in the consciousness of both men and women as also in the socio-cultural norms, the mass media and pattern of education all of which at present tend to perpetuate a passive unequal role of women in social, economic and political affairs. There is a need for a strong voluntary action involvement in order to evolve a specific strategy based on the local situation, in this area, of women. Any voluntary agency that is serious about promoting women's participation would have to seriously consider the challenge of recruiting and training women catalysts, extension agents and functionaries for reaching and eliciting women's participation.

Voluntary Action for Legal Aid: The majority of women have no knowledge about their rights and very few have resources to obtain legal redressal. It is now being strongly felt that laws by themselves cannot bring about the desirable change in the status of women unless women become aware of their rights. At the same time, it is also felt that since most of the women cannot afford legal representation in courts due to high financial costs and lack of access to knowledge pertaining to the legal process, the countrywide network of voluntary agencies can play an effective role in providing legal aid and legal education to women. One of the priority areas for the legal aid movement in the country should be the mobilisation of women through voluntary agencies. Voluntary organisations also have an important role in providing counselling, para-legal support and rehabilitation of women in distress. In fact, in the absence of such help, neither police nor courts can effectively help women. In many such cases, women are compelled to withdraw cases under the dowry prohibition act and compromise with unjust situations due to lack of alternatives. This situation could be remedied if women could be assisted through counselling support, employment training, and rehabilitation and development support by voluntary agencies.

The upsurge of interest in women's issues which characterised the decade, has left its mark on the legal scene. Voluntary organisations and activist groups are beginning to initiate action on various legal issues. It is felt that the Government should provide financial assistance to women's organisations for setting up legal aid cells. Evidently, there is a need for many more voluntary agencies to take up the issues of women and provide necessary legal aid to women.

Environment and Women: It is well recognised that the management of the environment requires the participation of people as they are closest to it and have a stake in its preservation. Active involvement of women and their organisations in environment protection is of paramount significance since women are most affected by the issue. There is a serious threat to the environment due to its degradation and pollution arising from various factors such as policies of government as well as the private sector, unplanned discharge of residual and waste, handling of toxic chemicals, indiscriminate construction of dams, large-scale deforestation, expansion of settlements and unplanned mining and quarrying work. Such conditions have pushed great number of women into marginal environment where floods, droughts, shortage of fuel, and excessive utilisation of grazing land have deprived women of their livelihood.

Many voluntary agencies have taken up the issue of environment protection. Among these agencies are the Dasholi Gram Swarajya Mandal that has started the Chipko Movement in which women play a very important role. This movement has received international acclaim and was initiated by hill women. Women embraced (Chipko) trees to prevent them being felled and some women were killed while thus protecting these with their own bodies. Trees to these women and others are the source of life. The impact of environmental degradation and of soil erosion is first felt by women. Many grass root level women's organisations have begun to take up environmental issues in addition to their continuing concern for rural poverty due to the intrinsic link among environment, poverty and gender.

The responsibility for creating awareness, mobilising public opinion and building a strong people's movement lies mainly with the voluntary agencies. The awareness generated by individual women pioneers/leaders and all types of women's organisations on environmental issues has focused on the fact that women and men have the capacity to manage their environment, and their access to productive resources should be sustained and enhanced. It is also recognised that the requisite knowledge and information can be disseminated by voluntary organisations to reinforce the self-help potential of women in conserving and improving the environment.

Demystification of Technology: A number of voluntary agencies are involved in commendable work in the, demystification of appropriate technology for the advancement of rural women.

The programmes implemented by voluntary agencies in this area include, providing opportunities for gainful employment and self-employment for women, reducing the drudgery in their lives, ensuring adequate medical and nutritional facilities, improving sanitation and environmental conditions and protecting women from occupational hazards. However, there is a need for further voluntary action in this direction that can develop and disseminate appropriate technology for women. Since the technological marginalisation of female work is endemic in both the agricultural and the non-agricultural informal sector, voluntary agencies should be involved in overcoming gender differentials in the application and generation of technology. There is a need to actively deploy technology to reduce the drudgery of the poorest working women in back breaking tasks such as gathering of fuel, fodder and water.

Training constitutes another important input, particularly for upgradation of skill and augmentation of earning capacities of women. There is a strong need for a diversification of training undertaking by voluntary organisations. Their role should be particularly geared to the sensitisation of administrators functionaries and catalysts on the issues and needs of women in development and in the delivery of comprehensive training programmes that have a component of knowledge, attitudes and skills for women's development.

Guiding Theories

The increase in the number, expansion and diversification of activities of the voluntary agencies has not necessarily equalised the disparities between them. There are not only regional imbalances in the growth of the voluntary sector but within a particular state, the growth of this sector has not been even. It is also well known that a majority of voluntary agencies are urban based and that relatively few have taken up the issues of women. It is recommended that the focus of voluntary agencies move from urban to rural areas, as the situation of rural women warrants immediate support. There is also need for the government to encourage voluntary action for the development of women by provision of adequate financial and structural support.

There is an urgent need to improve the effectiveness of voluntary action. Improvement will have to be brought about both in the organisational structures as also in the quality of services offered. Voluntary workers need to be professionals, equipped with appropriate skills for managing women's projects and sensitivity towards women's issues. National and State level institutions and training organisations should provide adequate facilities for research and training relating to women's issues. They should take up activities guided by the felt needs of women and those that can make a qualitative difference to women's lives, rather than be confined to traditional areas of support and action. Their work should be related to contemporary issues and thinking on women. As a caution, they should resist the temptation of initiating more work than they can effectively manage.

Voluntary action should be directed particularly towards preventive rather than purely curative measures. Efforts of voluntary agencies should also be geared towards generating self-reliance rather than to create dependencies. To improve their capabilities in planning and implementation of programmes, voluntary agencies are in need of expertise and technical guidance as much as financial assistance. Unfortunately, in the existing system of grants-in-aid, financial assistance to them assumes overriding importance vis-a-vis other forms of assistance such as technical guidance in the area of programme planning, project formulation, financial planning, administration, monitoring and evaluation. The proposed Resource Centre at the national level could also facilitate in the training needs of functionaries and in providing necessary managerial and technical assistance to voluntary agencies.

The process of grant-seeking and receiving is considered by a number of voluntary agencies as a frustrating experience. There is an urgent need to review the working of the grants-in-aid system. Wherever needed, modifications should be introduced to ensure that rules are simplified, grants released on-time and the amount provided is commensurate and proportionate to the needs of a particular programme. While a system of accountability for government funds is unavoidable, it need not be painful. Further, it should be ensured that financial assistance from the government does not seriously affect the basic character of voluntarism, its flexibility and innovativeness.

A number of programmes implemented by conventional voluntary agencies have emphasised imparting skills to improve the efficiency of women as housewives and mothers, and to improve their earning capacities. Voluntary agencies tend to neglect the participative potential of women in the development process as well as conscientising women on their rights and roles. There is the need for such efforts that could increase the awareness of women and improve their participation as equal citizens in national development. Further, voluntary agencies should play a surveillance role and observe, explore and analyse the extent to which social legislations implemented for women have actually benefited them. They should also act as pressure groups to better enforcement of laws for women. At present, there is no proper mechanism of coordination among different voluntary agencies working for the development of women. An effective mechanism for coordination between the government and the voluntary agencies is also to be ensured. CAPART and CSWB are appropriately situated to make efforts for more effective coordination and implementation of various programmes for women through voluntary organisations. For them to perform this role effectively, there should be a proper representation of grass roots women's organisations in CAPART and CSWB, that can serve as pressure groups. There is also the need for a continuous flow of information from government to voluntary agencies and vice versa through such mechanisms as a clearing house for information.

A focal point is desirable in the rural areas to encourage women's voluntary activities. Mahila Mandals and women's groups at the village or community level should be organised or revived and encouraged to register and function as women's institutions for undertaking socio-economic programmes. These institutions should be effectively linked with the various development and service agencies, offering training facilities for income generations as well as enhanced awareness among women. This linkage will enable women to absorb institutional finance for the development of viable economic activities. Particular attention will need to be given to the training of the Mahila Mandal functionaries and women's group organisers and provide them an orientation to development perspective rather than purely welfare approaches.

The CSWB which has been the coordinating agency for voluntary action for women and children, must respond to the new thrust of government policy meant for women and recast its own programmes. Greater coordination and cooperation among NGOs is called for to avoid duplication of services. Greater funding for networking among NGOs must be provided. This will ensure more efficient utilisation of funds and greater coverage of programmes. Government support to voluntary agencies for providing assistance to women in distress, including the running of crisis centres and short stay homes must be expanded. Para legal training must be an integral part of such efforts.

Voluntary agencies must be increasingly involved in the provision of employment and supportive services for women.

The National Literacy Mission must involve women's organisations in a big way. The voluntary sector should increasingly be involved to act as a catalyst/intermediary in organising women for collective action.

There is the need to document success stories of major NGOs in India and learn from their success and failures. Further, it is necessary to analyse the cost-benefit of NGO Projects versus governmental projects, i.e., both economic and social costs. It would also be critical to total the overall number of women reached by NGOs in India. The areas of activities and fields of success would also highlight their strengths and limitations.

In order to ensure that the security and integrity of the nation are preserved, there is a need to adopt suitable policies to ensure that voluntary agencies abide by the rules governing the receipt and utilisation of foreign grants and submit audited accounts, returns and reports periodically. Identity cards should be issued to workers of voluntary agencies who are dealing with cases of atrocities against women, as is already being done in some districts.

In order to have sufficient infrastructure and facilities, there is need to mobilise more resources for voluntary agencies which are engaged in welfare and development of women.

There is a need to decentralise the planning process to stimulate local people's participation in planning, implementation, monitoring and evaluation of development projects. A suitable mechanism should be evolved to involve voluntary agencies and other people's institutions at various stages of developmental programmes / projects. Voluntary agencies should further ensure the participation of poor women in the development process.

Recommendations have been made sectorally with a view to strengthening women's roles therein. Certain important issues, however, impinge on all spheres of women's lives and work. With a view to enhancing women's status and capacities to participate in the process of nation building, the following general recommendations are made:

The overall approach of this National Perspective Plan is to perceive women in a holistic manner. While the programmes for women will continue to be implemented by different ministries as part of their department plans, it is essential to have a strong inter-ministerial coordination and monitoring body along with its own supportive facilities service by the Department of Women and Child Development.

All ministries must reflect the concern for the all round development of women. The concerned ministries must have women's cell which currently only exists in the Ministries of Labour, Small-Scale Industry, Science and Technology and Rural Development. It is essential that the new policy thrust for women's development should be reflected in the Planning Commission as well as the State Planning Boards. The National Commission on Self-Employed Women and Women in the Informal Sector, has also independently concluded that the Planning Commission and State Planning Boards need to focus their attention sharply on the realistic situation of (labouring) women.

An essential prerequisite for the implementation of these new policy directives would be a women's unit in the Planning Commission, to redefine categories of data collection for women, modify existing terminology and identify gaps in data collection relating to women and to give direction to plans and programmes for women's development. It is also essential to analyse the impact of the different macro policies on women while planning new endeavours. Financial and fiscal resources should be apportioned and preferential allocations for women's employment in mainstream programmes and projects should be made. This would imply the rationalisation on resource allocation within mainstream programmes so as to benefit women, rather than only seeking separate allocations for women. Critical emphasis must be placed on rate of investment in women preferred industries and occupations.

At the State level, the Departments/ Directorates of Women's Development should be initiated. Currently, there is no separate department of women in many States. Social welfare, handicapped, Scheduled Castes and Scheduled Tribes are subjects that are bracketed together with the development of women at the State level. This new department could also be the State level implementation body for the programmes/ policies of the Department of Women and Child Development of the Government of India.

In terms of programme implementation, the two major implementing bodies envisaged, are the Social Welfare Boards and the Women's Development Corporations. There can be a rationalisation of service provision between these two bodies. The State Social Welfare Advisory Boards could eventually concentrate on implementing welfare/supportive programmes for women (homes for women in distress, working women's hostels, counselling centres for legal aid and paralegal training, condensed courses, etc.). Women Development Corporations would be responsible for the implementation of economic programmes through non-governmental and governmental agencies/ departments wherever necessary, concentrating on technical inputs like credit, marketing, design development, etc., and reaching out to women at the district and village levels.

Women should be entitled to a package of services at the block level created by the convergence of schemes such as Development of Women and Children in Rural Areas (DWCRA), Integrated Child Development Schemes (ICDS), Adult Education, Health Care, etc., at the grass roots administrative level. Every district should have a coordinator to assist in the integration of these programmes aimed at the development of women. The coordinator will also be responsible for motivating local planning of programmes and assist in their implementation and provide feedback for effective planning and evaluation. Since decentralisation of planning monitoring and implementation of development programmes for women is suggested as also devolution of finance at district level, appointment of District coordinators for women's programmes would facilitate this process, and control over finance would empower them. The National Commission on Self-Employed Women and Women in the Informal Sector has also recommended the appointment of District Coordination Officers to be responsible for planning, monitoring, coordination and evaluation of the programmes affecting women. Rationalisation of functionaries at the block and village levels to ensure coordination of programmes affecting women at the grass roots level also needs to be undertaken.

There are today sufficient number of programmes in the Government of India as well as innovative programmes in many States and sectors. What is needed is not merely larger resource allocation but technical inputs for greater effectiveness of these programmes, to guarantee better resource utilisation. Emphasis has to be placed on more effective planning, monitoring and evaluation of existing programmes through a result oriented mechanism operating at different levels.

Recognising that a critical input for women's development would be a new thrust to training and wider dissemination of information backed by research data and documentation. It is proposed to set up a National Resource Centre for Women. This resource centre would translate national developmental needs of women into a systematic grid of programmes and schemes for training at different levels in skills/ knowledge/ attitudes. The centre would identify and if necessary, strengthen existing governmental and non-governmental agencies including women's universities /women's centres and colleges through which the training, research/ dissemination could be carried out. The National Commission on Self-Employed Women and Women in Informal Sector has also recommended the need for a National Institute to cater to women's training as well as formulate guidelines and help the other constituent units at the State level, Divisional levels and district level to carry out training programmes.

Reorientation and sensitisation of the administrative machinery at all levels in the Government of India, the States, as well as specialised technical agencies (both Government and Voluntary) to the issues of women in development is essential. Three levels of orientation are necessary, i.e., at the policy and planning levels, at the district or intermediary level, and at the block and village levels. The training of functionaries and their orientation to women's issues must also be in the right perspective, i.e., women should be perceived as producers and participants, not clients for welfare. The dynamic role of women's contribution to the national economy as partners and equal citizens must be reiterated and translated into programmes and projects. The National Resource Centre would be responsible for revamping the existing content/ methodology and monitoring of training at all levels.

A special division should be created in the Department of Women and Child Development for the enforcement of law for women. The officer in charge may be designated Commissioner for Women's Rights and must liaise with the various Special Cells for women created by the police, the CBI as well as with the Departments of Public Grievances at Centre and State levels as also the Women's Cell in the Home Ministry. This division will be concerned with the enforcement of law to ensure women's rights, to facilitate action oriented research in needs such as discrimination against women, protection at work, etc. The Census in future must take into account women's unpaid work in the household and outside as well as the value added in performing her many survival tasks for the family. A greater conceptual clarity has to emerge on 'work' and 'non-work' as well as a distinction between work that produces economic value and other activities that are consumption oriented.

General Aspects and Approaches

The resilience in the social and cultural systems of the developing societies depends upon their balanced growth within and regulation of forces from without that influence their process of social transformation. Among the external forces the neo-colonial designs of 'big' nations assume a great deal of significance. An effective strategy of social and economic growth necessitates that nations in the developing world evolve a paradigm of modernisation which is commensurate with their culture, history and specificities of social structure. They may also have to work out a strategy of their relationship with 'big' power nations, specially with a view to safeguarding their cultural and political freedom. Such paradigm is necessary because modernisation inevitably activates social tensions, generates social inequalities and crisis of aspirations in the initial stages, in large segments of society.

Although the normative principles of modernisation for India have been enunciated in the Constitution, the paradigm of modernisation that India has adopted is reflected in its planning and development strategy. It comprises a consensual approach to development and modernisation, is based on principles of pluralism, decentralisation and distributive justice to be achieved through a rational legal path of democratic participation. It combines with state control over high points of economy with capitalistic mode of industrial and agrarian development. It calls for a strategy of development in which dissent and protest as democratic means of interest articulation are recognised as legitimate. These processes also help in the modernisation of society through dynamic and pluralistic consensus building among contending interest groups. It extends not only to the relationships among the classes, caste groups, or communities but also to the regions, states and centre and political parties in the regions. The consensual paradigm of modernisation adopted by India implies that social conflicts resulting from economic development require continual review of policies and management strategies of social change. The focus should continuously remain on twin strategies: first, growth and secondly, with growth distributive justice. The emphasis on protective discrimination for deprived sections of the people provided in the Constitution accords fully with this paradigm of modernisation.

A consensual paradigm of modernisation is indeed organic to the democratic process of development. The threats that most developing nations which adopt this paradigm encounter are both ideological and social structural. These threats are interrelated. The social structural forces give rise to counter-consensus ideologies and reinforce the forces of regression and conflict. Most developing countries witness a rise in the size of middle classes as economic development gathers momentum. Since Independence there has been a substantial growth in the middle class population in the social structure in India both in the countryside and cities. The green revolution in agriculture contributed to the rise of middle class, middle caste peasantry. In cities, a large segment of the entrepreneurial, professional and administrative personnel belong to the middle class. Most people in the services, education, journalism, media and public utilities jobs also give rise to the middle class-caste population.

The growth in the middle class structure has accentuated the crisis of aspiration among the rural and urban poor, the scheduled castes and the tribes. The condition of the poor has no doubt improved as a result of economic growth and modernisation. But the relative rate of improvement in the condition of the poor sections of society has remained comparatively slower. Hence a rise in their aspirations and social consciousness generates conflicts and social disaffection. The result is more tension, intercaste feud and communal violence.

The consensus ideology of modernisation, in operation since Independence, was designed to overcome such conflicts through social reforms, economic growth and distributive justice. The abolition of Zamindari, and ceiling on land holdings came soon after Independence. Later, the nationalisation of banks followed in order to release more resources for programmes of poverty removal and social welfare for the poor. By the 1970s most states also introduced protective discrimination in favour of the backward classes to improve their share in higher education, administrative jobs and in government sponsored programmes of economic growth and modernisation. These Policies effectively helped in conflict resolution and consensus building in most parts of the country, but such a process is an ongoing one and calls for changing strategy both qualitatively and quantitatively as the process of modernisation proceeds. The consensus, paradigm succeeded not only in social and economic domain of modernisation in India but also in stabilising the centre-state relationship, particularly in overcoming separatist movement of the fifties in the southern Indian states. The reorganisation of state boundaries on linguistic lines was a similar consensual strategy for political accommodation of regional, cultural and economic aspirations within the framework of a nation-state.

Ethnic Factors

When we evaluate the efficacy of the consensus paradigm of modernisation for building, resilience in the social system in the wake of forces of social transformation, we find that it has proved more effective in the management of social structural tensions than ideological ones. Consensus strategy has to a large extent contained the conflicts based on caste and class in the countryside and cities. It has been helpful in the management of inter-state, inter-political and centre-state relationships. It has maintained its legitimacy through policies and programmes of removal of poverty, of social welfare and protective discrimination.

In the ideological realm, however, consensus ideology in India has always remained on trial, as in this domain it had to reckon with communalism ideology promoted by the colonial rulers which resulted into the partition of India. The British colonial policy was systematically oriented to encouraging the ideology of communalism in India, be it of inter-caste, inter-tribal, caste-tribal or inter-religious nature. The consensus paradigm of modernisation which India adopted after Independence offered also a strategy for containing the communal challenges. The Indian leadership at no stage accepted the colonial communal ideology, the communal awards or the view that religious groups constitute separate nationalities in India. The 'two nation' theory based on religion was repugnant to national leadership as it was designed to weaken the forces of nationalism and the emergence of Indian nation-state.

Conscious of the threat that communal ideology posed to the balanced process of modernisation in India, the religious minorities have been provided many constitutional safeguards to protect and promote their cultural and religious identities. At the political level, the communal (religious) ideology of nation-state suffered a big set-back when Bangladesh broke away from Pakistan disproving the religious theory of nation-state. This established the validity of the secular ideology propounded during the Indian freedom movement as a basis for cultural and political modernisation. Nevertheless, the dimensions of communalism in India have assumed a multifaceted character, such as casteism, tribalism, regionalism and religious fundamentalism.

These communal forces have since Independence posed a continual ideological challenge to the paradigm of modernisation and the process of social and economic growth in India. During the early two to three decades after India gained freedom, the communal ideology derived its strength mainly from the exploitation of the economic, social and political deprivations of the minorities and deprived castes, tribes or religious groups. Communalism was used as a source for mystification of largely non-communal grievances.

To use a technical language, it represented a false consciousness. This variety of communalism could easily be taken care of through the consensus paradigm of modernisation of which the emphasis on growth with social justice was an organic component. Since the 1970s onwards communalism specially of the religious variety has taken a new form, that of religious fundamentalism. Unlike the former pattern, it does not, necessarily arise out of poverty or social or economic deprivation. Rather, it tends to grow with higher levels of economic development achieved by emergent new classes or groups within the minority religious communities.

Its focus is not merely on improving the economic, political and social life conditions of the religious group as such but on converting them to a counter modernisation ideology. It exhorts minorities for going back in time and postulates a religious worldview as a total alternative to the secular scientific worldview of a modern society. It postulates an overarching role for religion not only as an ethical system which governs personal faith, rituals and ways of life but also the economic, legal, political and social institutions. Religious fundamentalism does not believe in separation of politics from- religion, which is a challenge to the secular rational bases of cultural and political modernisation. It is not only regressive but also obscurantist in character.

The Indian social and political institutions have shown enough resilience despite many ups and downs in containing communalism of the false consciousness variety. Its modernisation paradigm has a built-in strategy to overcome strains generated by communalism. Partly it can be resolved through rapid economic growth and effective enforcement of the policies of distributive justice. The surveys on the changing economic and social conditions of the Muslims in various parts of the country reveal that all those among them who come from artisan occupational category have adjusted well to new entrepreneurial activities and have not only improved economic and social status but have also generated employment potential for others.

The institutional agencies such as the Minority Commission and Voluntary groups engaged in the economic and social welfare of the religious minorities, particularly the Muslims are steps towards containing the communal forces of the fundamentalist variety. These, however, pose a challenge to the resilience of the Indian social and political institutions as quite often such forces may be sponsored from outside the country. Communalism was used as a weapon by the imperial and colonial forces in the past, and there is no reason to believe that such forces might not be active again to retard the progress of autonomous growth of the Indian nation- state.

Its back-wash effect on the system, particularly the communalism of the majority community, the Hindus, may pose, in this context, a far more dangerous challenge than the communalism or fundamentalism of the minority communities. That Hinduism as a religion encourages values of tolerance, pluralism of faith and rituals or that it is a religion without a Church might harmonise with the secular ideology of modernisation but the potential for its communalisation in the contemporary historical context exists and it should be fought with all our determination. It calls for adequate institutional and political responses from our political parties and intellectuals.

As the process of economic development gathers momentum, the level of people's awareness of social, economic, political and ideological issues becomes sharp. This should by itself help reinforce institutional foci which have been created by the nation to overcome the forces of communalism and fundamentalism. However, additional institutional mechanisms may have to be devised. There is also a need for more effective monitoring and implementation of the programmes of distributive justice and, civic rights, specially of the minorities and weaker sections of society. The roles of mass media, education and voluntary agencies in institutionalisation of the Indian paradigm of modernisation can be termed as crucial. So far, despite continual challenges from communal forces, from within the country and without, India could adhere to its modernisation paradigm based on secularism, socialism and democracy. This has been possible because of the innate flexibility of this paradigm and also due to the strength of the Indian civilization.

The contemporary challenges, such as the rise of fundamentalism in Punjab and communal conflicts in other parts of the country can be seen largely as products of social frustrations generated by the contradictions of economic and social growth. For instance in Punjab the rise in communal violence is found to be related to the annihilative tendencies among peasant youth due to lack of employment opportunities commensurate with their rising aspirations. With higher levels of agricultural prosperity achieved by the Punjab farmers, aspirations of their younger generation have risen higher.

Agriculture no longer offers them satisfaction in terms of employment opportunities, especially of the kind aspired for the lack of rapid industrialisation reinforces this alienation further reinforcing the potential for violence. Most other forms of communal movements also owe their origin and legitimation to similar social and cultural conditions, although the combination of specific factors might vary. Adequate institutional responses together with more economic growth and industrialisation can help us meet the challenges of communalism and fundamentalism in India.

New Society

The cumulative result of the processes of modernisation in India since Independence have set into motion a massive process of social restructuration. Through this India enters into a new phase of modernisation. Its societal resilience is also put to new strains. The economic growth so far achieved has generated opportunities, enhanced social mobility, brought into dominant position a new middle class both in the cities and villages. The scale of professionalisation in the services, utility agencies, etc. has reached new high points. The successive electoral participation by the people, their exposure to mass media and the political, social and cultural movements have heightened their level of consciousness and non-formal education.

This has increased the cultural and motivational displacements among various classes, castes and communities. A great deal of structural inter-meshing has resulted among these groups based on the asymmetry of the principles of wealth, power and level of education. The community structure in India, be it of a caste, religious group, tribe or territorial group, no longer remains homogenous. The process of social mobility through new jobs, education, enterprises, access to political offices, etc. have severely fractured the homogeneity of communities, and made it possible now to look at the Indian structure in terms of categories, such as occupation, class, ideology, etc. rather than as communities such as caste, kinship, tribe or religious groups. The process of restructuration is increasingly eroding the community base of the social structure and slowly bringing into being alternate principles such as of class and category as elements of organisation of society.

The process is a natural result of investment in modernisation. But it also generates contradictions of serious proportion in the process of transition. No structural transition has ever been painless in history, but the democratic process of the Indian social transformation has been based on the premise that human cost of this transition must be minimised. The resilience of the system and its paradigm of modernisation can be tested in accordance with its success or failure in achieving this objective. How far is India able to achieve this goal? What are the sociological symptoms in this regard? Answer to these questions assumes significance for the Indian situation today, faced, as it is with gigantic phenomena of restructuration in society. The direction in which society is moving is indicated by the impact that modernisation forces are having in the creation of alternative social structure by fracturing the traditional ones which were based on caste by classes, professional groups and social movements in urban areas. Massive migration of the rural population to cities goes on alongwith this process. Many social reform movements are going on in India today which are organised not on the principle of caste, religion or community but on secular principles, such as the commonality of interests, skill and knowledge and commitment to social responsibilities. The social movements in the field of women's status, about ethicisation of professions, the legal aid movements and the voluntary groups for the development of the deprived sections belong to a new category altogether. The organising principle of such movements is not primordiality such as caste, kinship or community, but one's status as a citizen of the Indian republic. Many social movements in some parts of country aim at generating humanistic awareness among people about science and technology and its role in society. The people's movement about generating awareness of environmental problems created by unplanned industrialisation and its dysfunctions have also been increasing fast during the decade of the 1980s.

These developments augur well for realising the Indian goal of modernisation. Its impact, however, still remains limited in size and spread though strong in qualitative terms. The principles of caste, community and religion continue to articulate themselves as interest groups in the system. The significant change is the historical disjuncture today between the principles of community versus category in the processes of social restructuration in society which differs from the pattern of the 1960s and 1970s. In the decade of the 1980s the traditional institutions instead of wholly succeeding in adapting the modern institutions and values within their framework are slowly giving way to institutional replacements and differentiation.

This process indicates a qualitative shift in the pattern of social transformation in the Indian society. Despite the fact that it generates tensions or even occasional violence, it is a measure of assertion of the element of resilience in the Indian social system to cope up with processes of modernisation. In the new social formations, traditional institutions instead of absorbing the functions of the modern sector are most likely to undergo differentiation of forms without, however, fully eroding the traditional institutions and value orientations.

The nineteenth century, in India, may be noted for the most abominable conditions for women as well as the initiation of their emancipation. At the beginning of the century, the women were humiliated and tortured, kept illiterale, tradition bound and in complete subjugation to the men's will. As time passed, people began to rebel against this state of affairs.

The impact of western culture created a desire among many Indians to examine their beliefs and ideals. The scientific approach of the west initiated a search of Indian scriptures and mythologies for the causes of maladies that existed in society. A critical outlook among educated Indians aided the reinterpretation of religious texts and rational approaches replaced emotional reaction. These approaches led to the conclusion that blind faith, inability to discern right from wrong, and propagation of erroneous interpretations by decadent priestly class and immoral feudal lords had all been responsible for the existing web of subordination around women.

Regular Struggle

Reformers began their work in the direction of women's emancipation in the late nineteenth century, This work was continued in the twentieth century. In spite of great zeal and enthusiasm, the reformers achieved little success in breaking the strong notions of women's inferior status. However, the process of awakening to reality was initiated and India passed to an era in which myth around women began to be seriously questioned in terms of realities of human existence. Women started regaining their freedom of will and intellect. Still with the majority of women a major problem remained. They had for long believed that they were emotionally, morally, biologically and intellectually inferior to men and so were unable to contemplate their equal status with men. They failed to realise their full potential and high abilities. To alter this situation it was required that not only women, but the whole society should undergo a transition in its outlook towards them.

Let us now examine the movement for the emancipation of women, which was initiated in the nineteenth century.

It was largely through the efforts of Raja Ram Mohan Roy (1774-1833) that in 1829 Lord Bentinck decreed for the abolition of the sati system. He worked towards widow remarriages, inter-caste and inter-racial marriages, abolition of child marriages and polygamy. He founded a new religious order called the Brahm Samaj. This reformed order became well known for the privileges it gave to women.

Keshab Chandra Sen (1838-1884) became Roy's spiritual successor and worked tirelessly for women's education. The Civil Marriage Act III of 1872 raised the minimum age for marriage to fourteen, gave permission for widow remarriage and inter-caste marriage and penalised polygamy. This reform was applicable only to members of the Brahm Samaj and was brought about by Sen's courage and perseverance.

Many other prominent Indians were motivated by the example set by Keshab Chandra Sen. They led a vigorous struggle against child marriage and advocated widow remarriages. Ganga Ram at Lahore, along with others, started campaigns for women's emancipation. Mahadev Govinda Ranade (1842-1901) led a movement in Poona. He was a great patriot and social reformer who worked against existing Hindu customs, which made the life of the Indian women one of sorrow and hardship. It is notable that his reforms started with his own wife whom he taught himself.

In the north, Swami Dayanand Saraswati (1827-1883) founded a religious order named Arya Samaj. The basic tenet of this order advocated equality of opportunity for all irrespective of caste, colour, or sex. In a very active manner, this society worked for women's education. Dayanand broadened the scope of women's education by emphasising education for self rather than linking it to their efficiency in familial roles. The curriculum for education devised by Dayanand was similar for both boys and girls. Hence Dayanand emerged as the first reformer who advocated and evolved a comprehensive scheme of education for raising women's status not only at the level of family and society but for her own development as an individual.

Another prominent person of the nineteenth century, Ishwar Chandra Vidyasagar (1820-1891) helped the British Government to establish the first girl's school in Calcutta in 1849. He was further responsible for the establishment of forty girls' schools in Bengal between 1855-1858. The Government Act of 1856, which legalised widow remarriage for Hindu women, was passed largely because of his efforts.

A brilliant national leader Gopal Krishna Gokhale (1866-1915) worked relentlessly towards raising the status of women in Indian society. He founded "Servants of India Society" in 1905 and established as its main objective the education of women. Two religious leaders who helped in bringing about the renaissance in the Hindu religion were Ram Krishna Paramahansa (1833-1866) and Swami Vivekananda (1862-1902).

Pandita Ramabai (1858-1922) was the most prominent of women of the nineteenth century who fought the battle for her own and her sisters' emancipation. Her book *The High Class Hindu Woman* was published in 1888 and highlighted the burdensome life of the caste Hindu women. She made a fervent appeal to the American (men and women) to render help for the enlightenment of Indian women and society. The last words of her book state:

> "In the name of humanity, in the name of your sacred responsibilities as workers in the cause of humanity, summon you, true women and men of America, to bestow your help quickly, regardless of nation, caste or creed."

American admirers sent Ramabai Rs 6000 to aid a widows' home, Sharda Sadan (home of learning), which she opened.

Another woman closely associated with education for women was Ramabai Ranade (1865-1922) who was the wife of Justice Ranade. She established an educational institution called Seva Sadan. Anandibai Joshi (1865-1887) was the first Hindu woman to take a Degree of Doctor of Medicine in America. Because she died very young, she could not provide a much-needed influence on education in medicine for Indian women. However, her life and work inspired women of late nineteenth and early twentieth centuries. Francina Sorabji (1833-1907) with her daughters, particularly her fifth daughter Cornelia planned and organised the establishment of schools for girls in the Western region of India. Cornelia was awarded a Law Degree from England.

Gandhi's Role

Mohan Das Karamchand Gandhi, a shrewd politician and advocate of the righteous path, raised his voice against injustices to women of India in the 1920s. He took up many causes, including widow remarriage, and drew many women into his fight for freedom. His writings, speeches and private advocacy of the women's cause were attended to with great respect. In an article in a publication called *Young India* on November 18, 1926 he made it clear that if a man could remarry after the death of his spouse, a woman should also have this right: "What is considered desirable for man should be equally so for woman, and therefore, a widow should have the discretion as widowers about remarriage". In the same article he made the following comments:

> "I look to every youth in India to resolve not to marry a girl under sixteen. Let us tear down Purdah with one mighty effort. All that I have said about the wife applies equally to the husband-. She is a co-sharer with him of equal-rights and of equal duties."

Gandhi's approach towards women was idealistic. Though he was aware of their woes, he didn't think they had the same failings as men. He believed that their potential goodness incapacitated by years of bondage would, once freed, alleviate all the ills of society. He wrote that woman is sacrifice personified. His ideal of love was feminine as he identified that the greatest asset of a female was the ability to give motherly love. He saw the family as the nucleus of all Social Life and believed that women best functioned at home rather than in the workforce. He saw the ideal of women with strong characters tending their families with peace and harmony and leading the world in this direction. He addressed a meeting of mill workers in 1920 at Ahmedabad:

> "It is not for women to work in factories. They have plenty of work in their own homes. They should attend to the bringing up of their children. If women go to work, our social life will be ruined and moral standards will decline".

Although Gandhi's efforts favoured the traditional role for women, he believed they deserved equal recognition for a job well done, just as men received. Gandhi's struggle for women's rights was that of dignity and humanity, rather than one of role differentiations. Certainly, in the tradition bound society into which his ideas were received they were quite revolutionary. His zeal led people to a realisation of injustices to women and prepared men to be more realistic in their attitudes towards women.

However, Gandhi still maintained the goddess image on a pedestal while exploding the myths about the witch like nature of women. He believed in the basic goodness of human nature and propagated the ideal of complete man and complete woman, both able to offer unconditional love while working for a common goal.

Gandhi's influence helped women assert their will. It was due to his leadership that women were given rights equal to men in India's Constitution. The history of the women's participation in the freedom struggle in response to Gandhi's call is one of the most fascinating stories of the rise of the Indian female, The Indian women of the late nineteenth century and early twentieth century who had led a life of drudgery behind purdah finally emerged free of the web that had held them captive for centuries.

Great reformers like Gandhi drew attention to the goodness inherent in women and made an appeal for an understanding of their human need. Men became aware of the love and affection which women were capable of bestowing. Many women realised that they were not inferior beings and were proud of the key role they played in the survival of human race. An important Muslim reformer of the early twentieth century, Sir Sayyid Ahmad Khan, began a movement for the education of Muslims under the western influence. However, he excluded women from his scheme and firmly opposed their education. In contrast Molvi Nazir Ahmad and Hali were in favour of education of women. Perhaps, because of the influence of his mother on his early life Sir Sayyid saw no need for women's education other than home learning. He opposed starting schools for girls but emphasised the education of Muslim boys at Aligarh. On the other hand, Hali had his women characters reflect on the backward condition of Muslim community, the stagnation of vernacular learning, and the need for girls to be educated in order to fulfil their household and family duties. He did not go to extremes; however, the reforms he advocated were justified in terms of women's traditional role. In a magnificent poem called *Homage to Silence* Hali versified the role of women and their right to education.

Among the foreign women who came to India to serve the cause of women's emancipation, the names of Margaret Noble, later known as Sister Nivedita, Annie Besant and Margaret Cousins stand out. These women of Irish origin and former participants in Irish Home Rule agitation took up the cause of the downtrodden and underprivileged in India.

In 1917 the question of women's franchise was mooted. A deputation of women headed by Sarojini Naidu as spokesperson presented a memorandum to the British Secretary of State in 1919. The deputation demanded women's suffrage as well as increased educational and health facilities. The decision about the franchise was left to the Indian legislatures. With little opposition from these legislatures, thanks to the influence of the reformers, by 1929 women were enfranchised equally with men.

Although women were granted political rights equal to men, much reformative work was still needed to be done, as purdah and child marriages continued to be the dominant traditions. The laws of inheritance and divorce to assert their will legally restricted women. Illiteracy was predominant and so also the dowry customs. Women's organisations raised these issues for review. Enlightened Congress leaders like Nehru, Patel and Rajendra Prasad gave their full support for removing social and legal inequalities, which had been put upon women. Some of the women leaders who deserve acknowledgement for their efforts in achieving higher status for women in Indian society are chronologically: Annie Besant, Sarojini Naidu, Muthulakshmi Reddi and Raj Kumari Amrit Kaur.

Some interesting events occurred in 1929 when Gandhiji started Satyagraha, a freedom movement into which many women entered wholeheartedly. Mahatma's Dandi march which was in defiance to the salt law, had the full support of women. Within the next three years of this march, over five thousand women served terms of severe imprisonment suffered from lathi blows, cruelty, loss of livelihood and ill health. They picketed wine shops and shops that sold foreign clothes. They faced trials in law courts, suffered atrocities in prisons and were defamed for denouncing purdah and other evil social customs by the orthodox community leaders. Society came to realise, to some extent, the capacities of women as wilful responsible persons. Men and women began developing mutual respect. The prominent women of this period were neither an enigma nor a myth but were real people fighting for their independence. Sarojini Naidu, Kamla Nehru, Rukmani Lakhshmipathi, Hansa Mehta, Nalini Sen Gupta, Satyavati Devi, Miraben (Ms. Slade), Durgabai and Kuttimala Amma all contributed richly to the destruction of the many strands of the web, which had dictated that women were either goddesses or witches. The above mentioned women were compassionate people who worked very hard for the good of all and for the betterment of their country.

Political Disinformation

In the general elections of February 1939, nearly five million women voters participated enthusiastically in electioneering. The majority of their votes were cast for Congress. It became clear that women could responsibly vote and the doubts of the orthodox men that the enfranchisement of women would bring their country to doom were quickly dispelled. The women who contested on the Congress ticket were extraordinarily successful. In, one constituency, a woman elementary school teacher defeated the Vice-Chancellor of the University of her Province. In another, the wife of a doctor overthrew the President of a District Board who had represented the constituency in Parliament for twelve years. In thirty-nine years, India underwent such sweeping changes that by 1939, the political, educational and social position for women had risen so high that eight women were members of provincial and state legislatures. India ranked third among the nations of the world in regards to the political influence and position secured by its women. The United States and Russia were first and second respectively.

Viney Kirpal in her introduction to the edited book, *The Girl Child in 20th Century Indian Literature* writes about the depiction of the girl child in the literature of pre-independence period that "as in ancient literature, girl children are not presented as girl children. Their chronological age might place them as children but they appear in those works as 'miniature women-child brides, child wives, child widows'. The girls may be young in age but the responsibilities they are depicted to be taking up are those of grown up women." She further reports that "the period of adolescence and the transition to womanhood is like a blank page in these texts".

After Independence in 1947 most of the evils in Indian society were vigorously brought to the focus of attention of the nation. The principle of equality among the sexes was effectively put into law. Those ideas were constitutionally accepted by the Indian people who had been prevalent in the advanced nations for a long time. When the Constitution of India was framed in November 1949. Its Preamble stated that equality among the sexes was a fundamental right. Many other laws were enacted, intense efforts were made to remove the long-standing legal disabilities around women. It is worth noting that equal voting rights for women were given the very day the Indian Constitution was adopted.

Governments' Role

Now, the government offers incentives to women to enable them to obtain education that will remove sex discriminations. No woman can now be refused employment because of sex. At present quite a number of women hold jobs in top administrative and managerial services. Almost all the services are open to women who are involved in serious competition with men for the topmost positions in foreign services, police and engineering. They have also found employment in areas such as nuclear engineering, flying and paratroops.

The Hindu Marriage Act of 1955 gave women the legal right to divorce. The daughter who was completely excluded from the inheritance of property among Hindus after the Vedic period now has this right which was granted by The Hindu Succession Act of 1956. In 1961 the Dowry Prohibition Act was passed. The Government began the family planning incentive programme as a social welfare measure to release women from unlimited child bearing responsibilities.

In spite of these measures to ensure equality, most women of India are sill tradition bound. Cormack, Rama Mehta, Promilla Kapur, Goldstein and others who have studied the thinking and the outlook of modern Indian women through questionnaire and Interview techniques found that the women, even the highly educated ones, are still completely rooted in their families. The unmarried still depend on their fathers for advice and guidance as well as for the choice of their husbands. The married depend on their husbands for advice about their careers, lifestyles and future life plans. These women have identities, which have meaning only in terms of men in their lives. Independence as is found in Western women is lacking in most of Indian women who are willing to sacrifice their independence and identity for the family and for their men who get precedence over their own selves. Since society as a whole accepts this orientation, those who challenge this notion have a difficult time. Divorce is still a hated word in most of the Indian families. Nearly ninety per cent of all marriages among the middle classes are the arranged marriages. The choice of marriage partners depends on the parents.

The education of women is now considered as an essential feature of our national life. The point worth consideration is that: "should education strive to change this situation or strengthen these practices?" Many scholars are of the opinion that the stability of the Indian society is due to the prevalent practices while the exponents of the Women's Liberation Movement are very much against the undermining of women's position in any respect.

Social Justice by Education

Social Variations

Change in Society: Society is a web of social relationships. Hence, social change is a change in social relationship. It is the change in these which alone we shall regard as social change. What are social relationships? Social relationships include social processes, social patterns and social interactions. These include the mutual activities and relations of the various parts of society.

Jones: "Social change is a term used to describe variations or modifications of any aspect of social processes, social patterns, social interactions, or social organisation." Thus, social change is a change in the social organisation.

Davis: " By social change is meant only such alterations as secure in social organisation-that is, the structure and functions of society. Social change can be observed in every society."

Merril and Eldredge: " Social change means that large numbers of persons are engaging in activities that differ from those which they or their immediate (forefathers) engaged in some time before."

Society is composed of a vast and complex network of patterned human relationships in which all men participate. When human behaviour is in the process of modification, this is only another way of indicating that social change is occurring. Human society is constituted of human beings. Thus, whatever apparent alteration in the mutual behaviour between individuals takes place is a sign of social change. This fact of social change can be verified by glancing at the history of any society. Man is a dynamic being. Hence, society can never remain static. It undergoes constant variation.

Major Factors: What are the factors due to which social relationships undergo constant alterations? Some writers consider diffusion to be the main factor of social change while some other writers consider invention in a similar capacity. Actually, both diffusion and invention have a hand in the change which besets social relationships. Roughly, the main causes of social change are the Cultural, Technological, Biological, Population, Environment, Psychological and other factors.

Cultural Factors: The main cause of social change is the cultural factor. Changes in the culture are accompanied by social changes. Max Weber has proved this hypothesis by a comparative study of religious and economic institutions. Actually, no one can deny that changes and variations in culture inevitably influence social relationships. Culture gives speed and direction to social change and determines the limits beyond which social changes cannot occur. This, however, does not warrant the conclusion that there is no distinction between social and cultural change. Actually, the field of social change is limited in comparison with the field of cultural change.

Technological Factors: The technological factor has immense influence in social change. To quote Ogburn, "Technology changes society by changing our environment to which we in turn adapt. "This change is usually in the material environment and the adjustment we make to the changes often modifies customs and social institutions. In this way, the incessant increase in new machines and methods due to new discoveries has had a very great influence upon social relationships. The form of society is undergoing change as a result of the development and invention of electric, steam and petrol driven machines for production, the means of transport and communication and various mechanical appliances in everyday life. Even institutions, like family and marriage have not remained immune to the effect of these developments. The explicit effects of technological advancement are labour organisation, division of labour, specialisation,' high speed of life increase in production, etc. In the modern age technological factors are among the predominant causes of social changes.

Biological Factors: Biological factors too have some indirect influence upon social change. Among the biological factors is the qualitative aspect of the population related to heredity. The qualitative aspect of population is based upon powerful and great men and mutation. Hence, biological factors play a part in social change to that extent. In addition to this, the biological principles to natural selection and struggle for survival are constantly producing alterations in society.

Population Factors: Even changes in the quality and size of the population have an effect upon the social organisation as well as customs and traditions, institutions, associations, etc. Increase and decrease of population, a change in the ratio of men and women, young and old, have an effect upon social relationships. Decrease or increase in the population has an immediate effect upon economic institutions and associations. The ratio of men to women in a society affects marriage, family and the condition of women in society. In the same way, the birth and death rates, etc., also influence social change.

Environmental Factors: The geographists have emphasised the impact of geographical environment upon human society. Huntingdon has gone so far as to assert that an alteration in the climate is the sole cause of the evolution and devolution of civilizations and cultures. Even if these floods, earthquakes, excessive rain, drought, change of season, etc., have significant effect upon social relationships and these are modified by such natural occurrences.

Psychological Factors: Most sociologists regard psychological factors as important elements in social change. The cause of social changes is the psychology of man himself. Man is by nature a lover of change. He is always trying to discover new things in every sphere of his life and is always anxious for novel experiences. As a result of this tendency, the mores, traditions, customs, etc., of every human society are perpetually undergoing change. This does not mean that man always considers the new to be superior to the old. While he is always attending to what is new and unique, he also wants to preserve what is old. The form of social relationships is constantly changing in the process of interaction between these two tendencies. New customs and methods which replace the old traditional customs are being formed. Old traditions are respected but time demands change and adaptation to changing conditions. Change is the law of life. When changes do not occur at the appropriate time, revolutions take place, wars are fought, epidemics spread and changes are violently introduced.

Other Factors: In addition to the above-mentioned factors, another factor or social change is the appearance of new opinions and thoughts. For example, changes in the attitudes towards dowry, caste system, female education, etc., have resulted in widespread social variations and modifications. In fact, a majority of the social revolutions take place as a result of the evolution of new ways of thinking. Similarly, war is also a cause of social change because it influences the population, the economic situation and ratio of males to females, etc. In the same way, social change are introduced by the advent of great thinkers as Gandhi and Karl Marx, etc.

The form of each aspect of social life is being continually transformed due to the effect of the above mentioned factors which cause social change. New institutions and associations are being formed and destroyed in the social, economic, political, cultural, and verily, in all spheres. The form of family, marriage, state, religion, culture, educational system, and economic and social structures, is continually changing and transforming, as a result of which, a change occurs in the life of the individual and subsequently in his relations with others. To take an example, the result of social change can be well understood and realised by studying the history of the,objectives, structure, forms, importance and functions, etc., of the family from the early past to the present day. Similarly, all the changes and differences that can be seen between the tribal society and the present day society can be attributed to social change.

Significance of Education: The functions of education in the sphere of social change are outlined as follows:

Assistance in Adopting Social Change: Whenever some social change occurs, it results in changes in the pattern of doing some work. This change is easily adopted by some people, while others find it very difficult to adjust themselves to this change. It is the aim of education to make all good changes easily accessible to every person. In India, a large majority of people found it difficult to reconcile themselves to the changes that occurred in the institutions of family and marriage, but the educated minority soon realised the advantages of these changes. Later on, as education spread to other sections of society, the importance and value of these changes came to be recognised more universally. For this reason, it is now generally accepted that before bringing about any social change in society, it is necessary to create a receptive temper of mind among the people. Otherwise, there is invariably some resistance to change.

Overcoming Resistance to Change: Certain factors make it easy for some social innovation to be adopted and accepted, but on the other hand, certain factors create resistance to acceptance. The best way of overcoming such resistance is education. Through the medium of education, the importance of social change is convincingly explained to the people so that they are purged of their prejudices and blind faith, and this enabled and also strengthened to accept something new.

Analysis of Change: But society does not progress with any and every change, and neither does the individual. Progress occurs only when the change leads the society to more desirable social values. And for this, analysis and criticism of social change is essential. Only the educated individual can make valid criticisms and offer constructive suggestions, because it is education which invests the individual with the capacity to use his intelligence, to distinguish between right and wrong and to establish certain ideals. Education determines the values which act as a criterion for the analysis of social change. Through this analysis and criticism, undesirable social changes are prevented and desirable social changes encouraged. In every society, this is achieved only through the efforts of rational and educated people.

Emergence of New Change: Since the educated class is constantly engaged in an analysis of contemporary society, it also makes frequent suggestions for improvement. It is on the basis of such suggestions that social reform movements are set into motion. The flood of social reform movements that was witnessed at the turn of the century was due to modern education. Educated people made a study of western societies and their institutions compared both with our own, and agitates public opinion towards the abolition of many social evils such as child marriage, objections to widow remarriage, unequal rights of women, the custom of women committing *sati,* etc. Only through education could the importance of such changes be made apparent to the people. It was through education alone that public opinion could be turned in favour of these changes. Hence, it is the educated class in every society which intimates, guides and controls movements for social reform.

Leadership in Social Change: If social change is to be directed properly, it is necessary to have able leadership, well acquainted with the complexity of the problem. Such leaders can be created only through education. For this reason social welfare workers must first be educated before they are unleashed on society. On the one hand, education creates in them a consciousness of social change, while on the other, it teaches them to distinguish between the good and the bad. Education in India must be able to create appropriate leadership at every level if social changes conducive to democracy are to be introduced.

Educating the People of Social Change: Only a properly organised system of education can generate in the people of a society the ability to adopt some social change. As a result of education, individuals learn to analyse their customs and traditions, to criticise them, and to cooperate in movements for social reform. The educational system has played a major role in bringing about revolutionary social changes in England, France, America and Russia.

Advances in the Sphere of Knowledge: New researches and inventions all depend upon education, because only the educated individual can search for new things in every sphere. Only such people can help in the progress of non-material culture. Fresh discoveries in the sphere of knowledge provide the right basis for criticism of society. Then the need for change becomes apparent. This, education contributes to social change by bringing changes in knowledge.

It is evident from the above account that educationists, educators and schools have a tremendous responsibility in social changes. Defective education leads to defective social changes. Hence, if society is to change in the right direction, it is essential that attention be paid to the educational system because this alone can create future generation of critics of change and leaders of reform.

Compulsory Education

In 1950 the Indian Constitution provided that all States should provide free and compulsory education to the children of every section of the society up to the age of 14 years in ten years time. As the literacy then was only 16 per cent in India, the decision was timely important and according to the need of the land. The success and prosperity of a democratic system of government depends on enlightened and educated people. Equal opportunity to all is the salient feature of a democratic set up. Hence, it is necessary that all people should get education. There should be no discrimination between rich and poor, touchables and non-touchables, low caste or high caste, Hindus and Muslims, Sikhs or Christians, etc. People become aware of their duties and responsibilities, rights and obligations only through education. Knowledge generates in them the feelings of nationalism, partriotism and sacrifice. Education changes people's behaviour for better. They become civilized, and learn and practise noble behaviour towards others.

Purposes and Goals

1. Free distribution of midday meals to the poor and needy children.

2. Supply of free textbooks and clothing to the poor children.
3. Directives may be issued by the centre to backward States to concentrate on the clearance of the backlog of non-attending boys and girls.
4. Setting up of School Improvement Committee for undertaking intensive drive for bringing not-attending children to schools and also to see to it that the enrolled children are retained in schools. Where School Management Committees already exist, they may be entrusted with this task.
5. Steps may be taken to enforce attendance, at least to the extent of issuing warning notices and attendance orders to the parents of defaulting children.
6. Whenever necessary, the prescribed teacher-pupil ratio may be relaxed while sanctioning new schools and additional teacher units in backward areas.
7. Provision of part-time schooling may be arranged for those children who are unable to attend regular schools.
8. Intensifying social education programmes in backward areas and among socially backward groups of people for educating the parents.
9. Special targets may be fixed for enrolment by the State Government from year to year for each district, greater attention being paid to backward districts and areas.
10. Separate targets for the enrolment of children of groups of backward classes may also be fixed at State and District levels.
11. It may be made obligatory for teachers to stay in the vicinity of the school as far as possible. As an incentive, payment of rural accommodation allowance to teachers of rural areas who live within the vicinity of the school may be considered.
12. Suitable facilities may be given to the children of rural elementary school teachers studying in high schools.
13. Residential type of schools (Ashram Schools) may be established for children of teachers working in very backward areas and the full cost of their education may be borne by the Government.

Improving Enrolment: Concerted efforts should be made for improving enrolment at the elementary stage. The following measures can help:

***A System of Multiple Entry*:** It should be introduced so that children could be admitted at different points. This flexible arrangement will keep their interest alive as they can be admitted to the class for which they are fit.

Part-time Classes: There are more dropouts at the age of nine, and this is due to economic reasons. If part-time classes are organised, many children may get education at the time most convenient for them. The duration and timings of these classes should be determined in the light of local conditions.

Own-time Education: This means self-instruction through special workbooks prepared for the purpose. This method can work well if some teachers are put in charge of some areas to help students in needs.

Provision of Incentives: A variety of incentives should be provided, like free distribution of textbooks and stationery, midday meals, uniforms and attendance scholarship. For promoting education in the tribal areas, and increasing number of Ashram schools should be established. Special training should be given to teachers in the tribal life and culture, provision of audio-visual aids, suitable curriculum, economic incentives to parents, etc., should be made.

Improving Quality

Broad-based Education: The objective of teaching in the primary schools should be the 3 R's good manners, healthy habits, some skill with hands, general knowledge and develop qualities like sense of responsibility, cooperativeness, discipline and patriotism.

Reformed Curriculum: Through the adoption of suitable curricula and appropriate teaching methodologies, the schools should inculcate in the students, such qualities as are relevant to the entire spectrum of occupations and would ultimately improve their adjustability and employability and make them more capable of setting down in self-employment, work-experience should form an integral part of the curriculum.

Improvement in Buildings and Equipment: School building should be constructed with the help of the local community. School should be provided with science- kits, radio sets. TV, etc.

Experimental Schools: Experimental schools should be set up to improve the content and methodology of teaching. NCERT, SCERT, Colleges of Education can take up these projects.

Better Status of the Teacher: Only trained teachers should be appointed and in-service training must be systematically provided. Social and economic status of teachers should be improved. There must be scope for further promotion.

Reform in Educational Administration: The administration should be decentralised as far as possible. The inspecting authorities should provide leadership to the schools.

Adequate Financial Provision: The Central and State Government should ensure that primary education is given adequate share out of the total outlay on education.

Local Resources to Help Primary Teachers: For example, if there is a good local singer available, his service may be utilised for teaching music to children.

Research: Research programmes on various aspects of primary education should be taken up for the improvement of primary education, *e.g.*, wastage and stagnation, curriculum, methods of teaching, action, research, etc.

The causes for shortfalls in the enrolment of girls at the primary and middle stages of education are manifold.

Economic Reasons: Mothers are reluctant to send their daughters to schools as they are very useful at home for carrying out domestic duties. Undernourishment and inadequate clothing and books in rural area is yet another reason. Hence, free uniforms and free books to the needy and deserving children should be given. Attendance scholarships should also be given as a compensation to parents. Midday meals should be provided.

Social Customs: Certain harmful social customs, such as purdah system, caste barriers, etc., stand in the way of girls education. Many parents who like to educate their girls are unable to do so because of co-education . To remedy these problems, research in women's education should be taken up, separate schools for girls should be established at Middle and High school stages, creating public opinion in favour of girls' education. Mass media, PTA may also be of help.

Lack of Girls Schools: One of the factors responsible for lower enrolment of girls is non-availability of a school in rural localities and lack of separate sanitary facilities for girls in mixed schools, lack of suitable school buildings and equipment. The following measures would overcome these difficulties.

1. One primary school within a radius of one mile from every home.
2. Hostel for girls.
3. Stipends to girls who are residing in hostels.
4. Transport facilities.
5. Free education for girls up to SSLC Examination.

Lack of Qualified Women Teachers: At present the proportion of women teachers to men teachers is very low. In Kerala it is 45 per cent, Tamil Nadu 33 per cent, Karnataka 25 per cent, and West Bengal 14 per cent. In the backward States the positions is much worse. It is 5 per cent in Orissa and 10 per cent in Rajasthan. The following steps may be taken to increase the number of women teachers: i) Special training institutions for women must be started and they must be located in rural areas, ii) Condensed course should be started to enable unqualified women for employment as teachers, iii) In rural areas, quarters for women teachers should be provided, iv) Rural allowance should be given to women teachers who are employed in rural areas, v) In-service training must be given systematically.

Lack of Lady Teachers: The number of lady officers is far too small, to shoulder the responsibility of speeding up the progress of girls education as envisaged in our plans. The offices are poorly staffed and ill-equipped. They lack conveyance facilities. To overcome these problems: a) Women inspecting officers should be increased, b) Adequate transport facilities should be provided, c) Adequate office staff and equipment must be available, and d) Residential facilities should be provided to all woman officers.

Local Schemes: The District Primary Education Programme (DPEP) is a special thrust and a new initiative to achieve Universalisation of Elementary Education (UEE). The programme takes a holistic view of primary education development and seeks to operationalise the strategy of UEE through district specific planning with emphasis on decentralised management, participatory processes, empowerment and capacity building at all levels. The programme is implemented through the State level registered societies.

The programme aims at providing access to primary education for all children; reducing primary dropout rates to less than 10 per cent, increasing learning achievement of primary school students by 25 per cent, and reducing the gender and social gap to less than five per cent.

The programme is structured to provide additional inputs over and above the Central/State sector schemes for elementary education. The programme fills in the existing gaps in the development of primary education and seeks to revitalise the existing system. DPEP is contextual and has a marked gender focus. The programme components include construction of classrooms and new schools, opening of non-formal/alternative schooling centres, appointment of new teachers, setting up early childhood education centres, strengthening of State Councils of Educational Research and Training (SCERTs)/District Institute of Educational Training (DIETs), setting up of Block Resource Centres/ Cluster Resource Centres, teachers training, development of teaching learning material, research based interventions, special interventions for education of girls education, SC/ST, etc. A new initiative of providing integrated education to disabled children and distance education for teachers training has also been incorporated in the DPEP Scheme.

The district is the unit of programme implementation and selected on the basis of twin criteria, viz.: a) educationally backward districts with female literacy below the national average and b) districts where Total Literacy Campaigns (TLCs) have been successful, leading to enhanced demand for elementary education.

DPEP is a Centrally-sponsored scheme. Eighty-five per cent of the project cost is shared by Government of India (GOI) and

15 per cent by the concerned State Government. Both the Central share and State share are passed on to State Implementation Societies directly as grant. The GOI share is resourced by external funding. Several bilateral and multilateral agencies are providing financial assistance for the DPEP. The World Bank has provided a credit amounting to US $ 260 million (approx. Rs 806 crore) under phase-1 of DPEP (1994-2001). The European Community has signed a financial agreement with the Government of India to provide a grant of 150 million ECU (approx. Rs 585 crore) as programme support for DPEP in Madhya Pradesh (1994-99). An agreement has been signed with IDA for a second credit amounting to US$ 425 million (Rs. 1,480 crore) for DPEP- 11(1996-2002). The Government of Netherlands has provided a grant of US $ 25.8 million (Rs. 90 crore) for DPEP in Gujarat. A grant of $ 42.5 million (Rs. 220 crore) for DPEP in Andhra Pradesh and $31.7 million (Rs. 207 crore) for DPEP in West Bengal is available from ODA (now DFID), UK. IDA credit of US $ 152.4 million (Rs. 530 crore) and a grant of US $ 10 million (Rs. 36 crore) from UNICEF has been tied up for phase-Ill for DPEP in 27 Districts of Bihar. World Bank has also offered further assistance of US $ 119 million (Rs. 452 crore) for expansion of DPEP in 14 additional district of Andhra Pradesh and US$ 220 million (Rs. 800 crore) for covering 19 districts of Rajasthan under DPEP. The Government of Netherlands has also made an offer of US $ 25 million (Rs. 100 crore) for covering 3 additional districts of Uttar Pradesh under DPEP.

The programme, which was initially launched in 42 districts of seven States in 1994, is now covering a total of 149 districts in 14 States. Expansion of DPEP in 14 additional district of Andhra Pradesh, three additional districts in UP and 19 districts of Rajasthan is in the pipeline.

The programme implementation has shown promising results as per reports of various supervision missions and evaluatory studies. The first in-depth review of DPEP was conducted between 26 September and 16 October 1997 through a mission comprising representatives from World Bank, European Community, Overseas Development Agency (UK), UNICEF, Netherlands and Government of India. The Mission has observed that the impact of the DPEP on the entire-primary education system is being evidenced. There has been significant increase in enrolment in the DPEP districts as compared to non-DPEP districts, gender and social inequities has been substantially reduced and preliminary signs of increased learning achievements are in evidence.

Fundamental Difficulties

J.C. Aggarwal, in his book 'Development and Planning of Modern Education' with reference to India, has listed the following major problems of universalisation of education:

1. Uneven spread of education.
2. Low enrolment of the backward section of the society.
3. Stagnation.
4. Wastage.
5. Low enrolment of girls.
6. Apathy and poverty of parents.
7. Defective curriculum.
8. Uninspiring methods of teaching.
9. Lack of reading and writing material for children.
10. Lack of qualified teachers.
11. Frequent transfer of teachers.
12. Lack of effective inspection and academic guidance by the inspecting staff.
13. Faliure to enforce compulsory attendance.
14. Lack of suitable admission policy.
15. Conservative attitude towards co-education.
16. Inadequate and unattractive school building.
17. Poor nutrition.
18. Existence of large number of incomplete primary schools.
19. Lack of part-time facilities.
20. Group involving local bodies.
21. Meagre financial outlay.
22. Overpopulation.

Inequality in Education: The total number of non-enrolled children at the elementary stage (classes I to VIII) is of the order of 470 lakhs. Most of them belong to the weaker sections of the country like Scheduled Caste, Scheduled Tribe, Agricultural landless labourers and urban slum-dwellers. Even in states, where overall enrolment has been satisfactory, low enrolment was observed in respect of girls, Scheduled Caste, Scheduled Tribe and in backward areas. In certain States, the very low enrolment of girls had become a problem. In all States some districts were advanced, whereas other were lagging behind. Similarly in the same district, there were differences in the provision of educational opportunities in different Community Development Blocks. The pattern of inequality of educational opportunity may be considered at several levels and with reference to different sections of society as follows:

1. Inequality between one State and another.
2. In a State, the prevailing inequality between one district and another.
3. In a district, inequal educational opportunity in different areas.
4. Inequality of educational opportunity between boys and girls.

Inequality of educational opportunity in India prevails between the different sections of society, advanced castes Vs. Scheduled Caste and Scheduled Tribes, upper and middle classes Vs lower classes, economically better off classes Vs poorer sections, etc. On the basis of the experience reported in the different states, the following causes of inequality may be listed:

1. Some States are economically advanced while others are lagging behind. Consequently, the income per head of population in different States varies considerably. The same is true of district, block and local level.
2. Social and psychological reasons, *e.g.*, apathy towards girls' education, particularly in socially backward groups of people.
3. Varying literacy levels in States, districts and localities.
4. Existence of inaccessible and isolated small habitations, particularly in hilly and forest areas.
5. Varying occupational opportunities prevailing in different areas.

6. Lack of suitable and adequate accommodation for running schools.
7. Dearth of suitably qualified teachers, particularly women teachers and teachers for tribal areas.

Opportunity in Education: The financing of Elementary Education should be separated from financing of other sectors of education and treated on a special footing. Special financial assistance should be given to all States on the Principle of Equalisation in order to enable them to fulfil the directive of Article 45 of the Constitution. While adopting this principle the Equalisation Authority should consider both development and committed expenditure on elementary education. The extent of State effort and the quantum of assistance from the Centre should both be decided by the Equalisation Authority, *i.e.*, the Central Government while equalising at the State level. Similar principles should be adopted by the State when equalising at the District level and by the District, when equalising at the local level.

Social Features

The following things can be stressed in this connection:

Knowledge of Rights and Duties: If democracy is to be a success, it is essential that every citizen should be aware of his rights and duties because only then can he take active and productive part in the affairs of the state. This knowledge of rights and duties can be obtained only through education. Education socialises the individual so that he develops consciousness of duty.

Development of Humane Qualities: If the ideal of brotherhood is to be achieved by a democratic state, it is necessary for it to develop humane qualities in its members. Kant's moral concepts throw important light on this. Only through education can such qualities as a high moral character, sociability, benevolence, patience, pity, sympathy and brotherhood, etc., be developed in the individual.

Faith in Democratic Ideals: In order to make democracy a success, it is essential that its citizens must have faith in the democratic ideals. And this can be brought about only when they are adequately educated because it is only the educated person who realises that the sole purpose of life is not the satisfaction of gross physical desires. The ideals of freedom, liberty, brotherhood are more valuable and necessary. No one, but the educated individual can understand the circumstances and needs of another person before passing judgement on him. Only such a person can accept the idea of equality after recognising human values as being the end to be achieved.

Fulfilment of Political Duties: In a democracy the government is elected by the people, and hence the responsibility for electing a good government develops upon them. And, if the people are unable to understand their political rights or to fulfil their political responsibilities, it is foolish to hope for a democratic government. This ability to recognise where one's duty lies can come only through education. Educated people can properly assess the qualities and shortcomings of the various individuals who are fighting the elections, and of the various political parties and their plans and policies which they profess. In India, the absence of education is a big handicap in educating a truly democratic state because during elections, the ignorant people are persuaded to vote for the wrong persons, with the result that the governance of the country has failed time and again. Corruption is rampant. The Mudaliar Report points out that if democracy is anything more than voting blindly, then every individual must accept the task of independently thinking about all social, political and economic problems before deciding upon the party he wishes to support. But this is possible only when the entire electorate is educated to think independently.

Protection and Transmission of Culture: In any state, ideals can be achieved only when change is accompanied by a parallel continuity, and this continuity with the past is maintained only through culture, the social heritage, which is passed on to the new generation through the medium of education. Hence, education is also required for transmitting culture to future generations and for protecting it.

Preventing Exploitation: The ideal of democracy are opposed to exploitation of every kind, but if political, social and economic exploitation is to be eliminated from society, it is essential to have universal and compulsory education. In its absence, the rich and powerful people will never give up their advantage and habit, while the poor will never become sufficiently conscious of their rights or their ability to organise together and counter this exploitation. Educated people in a country are aware of their rights and they have the intelligence and training to fight exploitation or violation of their rights. Hence, education is the only real foundation on which democracy can be based.

Part of Democracy

All these important aspects of democratic education hold true in India also. India is not merely a modern democratic state but a country which is traditionally inclined towards democracy. A democratic Constitution was adopted after Independence. In 1938, Jawaharlal Nehru had said, "The Indian National Congress stands for independence and a democratic state". This objective was achieved after Independence with the establishment of a democratic society. The Indian Constitution seeks to establish a popular government in the country on the basis of democratic principles outlined earlier. For this every citizen must participate in the administration, through his right to vote and to be elected. Every individual is guaranteed and given equal status and opportunity, because no one is discriminated against on the basis of religion, race, caste, community, sex, or on any other grounds. The government is responsible to the people and its elected representatives.

In order to achieve this objective of democracy, education is as necessary in India as anywhere else, a truth which the Indian people have been quick to realise. In the words of Dr. F. W. Thames, "Education is no exotic in India. There has been no country where the love of learning had so early an origin or has exercised so lasting and powerful an influence. From the simple poet of the Vedic age to the Bengali philosopher of the present-day, there has been an uninterrupted succession of teachers and scholars." Not only did the Indian Constitution accept the ideals of democracy, it considered education the prime responsibility of the state. In Article 45 of the Constitution, it has been stated that every state must arrange for the provision of free and compulsory education to all children up to the age of 14, within ten years of the date of inception of the Constitution. After the achievement of independence, a new phase began in the history of education. Articles 29 and 30 of the Constitution give fundamental rights to every individual in connection with education and cultural development. According to Article 29, every Indian national living in any part of India will have the right to maintain his own specific language, script and his culture. No person can be refused right of admission to any educational institution, established by the state, by reason of religion, race, caste, language or any other similar consideration.

According to Article 30, every minority community will have the right to establish and maintain educational institutions of its own choice, irrespective of whether the minority is a linguistic or religious one. The state will also not refuse aid to any such institution created by a religious or linguistic minority. Articles 45 and 46 determine the policy for education as part and parcel of the directive principles. According to Article 45, the state will make efforts to provide free and compulsory education, within ten years, to every child below the age of 14 .

According to Article 46, the state will pay special attention to the educational and economic interests of all backward classes, especially the scheduled castes and scheduled tribes. It also entrusts the state with the duty of protecting backward classes from social injustice and exploitation of every kind. The Indian Constitution laid the foundation for a federal government in which the functions of the state and central government are distinctly defined. Both the central and state governments have some duties with respect to the education. It has been realised that there must be coordination between the central and state authority on education for a balanced development of the country. The modern Indian state is a welfare state whose objective is the complete development of its people. This welfare can be achieved only through education. Little surprise therefore, if all the leaders of the nation stress the importance of education as a first step to improve the future of the nation.

Education for Liberty

In last decades, along with the growth of population, there has been a shift in the population from rural to urban areas. The increasing urbanisation has led to problems like crime and juvenile delinquency, alcoholism and drug abuse, housing shortage, overcrowding and slums, unemployment and poverty, pollution and noise, and communication and traffic control among others. But if cities are places of tensions and strain, they are also the centres of civilization and culture. They are active, innovative, and alive. They provide opportunities to achieve one's aspirations. If the future of our country is linked with the development of rural areas, it is equally linked with the growth of cities and metropolitan areas. But before analysing these problems, let us understand the basic concepts.

System Improving

One important solution to our urban problems is the systematic development of the fast growing urban centres and planning an investment programme which, over the next 20 years or so, could give rise to a large number of well distributed, viable urban centres throughout the country. So far we have been focusing attention on programmes for providing wage employment in rural areas through IRDP, NREP and JRY to hold people back in the villages.

While there is ample justification for providing rural employment, this by itself is not enough. It is not possible to provide gainful employment in the agricultural sector beyond a certain point. For this purpose, we have to emphasise on programmes which can permit multifunctional activities to sustain people in cities.

Process of Planning

Urban planning is almost city-centred. We have always been talking of town and city planning but never of the planned development of the whole region so that population is logically dispersed and activities are properly distributed. City planning is an ad-hoc solution but regional planning could be a more lasting one. For example, instead of providing houses to slum-dwellers in cities through city development authorities if through regional planning migrants could be diverted to other areas which may provide attractive employment, the pace of growth of existing cities could be checked. It is time that at least beginning from the Ninth Five Year Plan (1997), the Government of India helped the states in setting up regional planning organisations and evolving meaningful regional settlement plans.

Encouraging Industries to Move to Backward Areas: Land pricing policy which gives land in large chunks at throwaway prices has to be replanned to encourage industries to move to backward areas/districts. This will also take care of linear development of metropolitan and big cities. A policy of the state taking over potential high value land in and around large cities with a view to exploiting its full cost at a later date also needs serious consideration.

Municipalities to Find Own Financial Resources: People do not mind paying taxes to the municipality if their money is properly utilised to maintain roads, provide sewage system, reduce water shortage and provide electricity. It is a well known fact that cities suffer from crippling resource constraints. If deterrent punishment is given to the corrupt municipality officials, there is no reason why the municipal corporations should find it difficult in collecting money from the residents of the city. A city must bear the cost of its own development. High financial support from state government is becoming difficult. By revising property, water and electricity taxes, money can be collected and more money per head per annum can be made available for providing necessary amenities. When any new industry or business is located in a city or on its periphery, it could be moderately taxed so that additional money becomes available to the local body.

Encouraging Private Transport: Why should city transport be a public monopoly? When the transport is handled by state employees, it has been noticed that they tend to have extremely rudely and callously. Backing of the trade union encourages them to go on strikes frequently. It is necessary then that private transportation be encouraged. Privately operated mini bus and tempo services will charge a little more fare but commuters would not mind paying this in view of the better services.

Amendment of Rent Control Acts: Laws which inhibit the construction of new houses or giving of houses on rent must be amended. Which landlord would like to spend Rs 3 or 4 lakh or so on a two-three room tenement and give it on rent for Rs 500-700 a month or so for the next 10 to 20 years without having the authority to increase the rent or get it vacated on appropriate grounds. Maharashtra has taken a lead in amending the Rent Control Act which has made thousands of houses available for rent. A similar step in other states would be welcome.

Adopting Pragmatic Housing Policy: In May 1988, the central government presented the National Housing Policy (NHP) to the Parliament which aimed at abolishing 'homelessness' by the turn of the century and upgrading the quality of accommodation to a fixed minimum standard. Such policy sounds too ambitious and Utopian. It is a dream impossible to accomplish in a span of remaining four years by which time the twentieth century will have become past. The government policy and planning has to be more down to earth. This is not to say that the concept of NHP is irrational. The NHP strategy is broadbased. It seeks to provide easy access to finance as well as land and materials for building houses at reasonable rates. It also seeks to encourage manufacturers to use new type of building materials. Moreover, it seeks to review the entire gamut of laws relating to land tenure, land acquisition and ceiling to apartment ownership, municipal regulations and rental laws. But these are all thorny issues. The NHP is oriented towards rich developers, landlords and contractors. The NHP has to discourage luxury housing and promote cooperative and group housing societies. It has to develop special schemes for the poor and low-income people. It has also to favour providing incentives to employers to build houses for the employees. It has to increase its authorised capital of Rs 100 crore which cannot go anywhere near to meeting the financial needs. Unless, a more pragmatic NHP is adopted, it will be impossible to achieve the set goals.

Decentralisation of Authority and Responsibility: One proposal by innovative planners and some radicals envision a structural decentralisation of local self government itself. This could entail the creation of neighbourhood-action groups, to be called 'community centres' consisting of representatives of residents and municipality officials. These centres will identify and act upon neighbourhood needs. For example, many new colonies have come to be established in many cities in which as many as 10,000 to 50,000 people reside. Thus, these colonies are small towns by themselves. Some taxes like house tax, road tax, light tax, etc., could be passed on directly to these community-centres instead of giving them to municipalities. The centres would direct the affairs of the neighbourhood without reference to the city municipal corporation and use the collected money in maintaining roads, lights and so forth. The argument for this kind of decentralised structure within the city is that the same system that allows lakhs of people a substantial control over their civic destiny denies them an effective role in shaping the institutions that shape their lives. Community centres will allow them to create their own exclusive environment.

To conclude, it may be pointed out that the effects of urbanisation and urbanism and the problems of cities can never be solved until urban planning is modified and radical measures are taken. These should not be based on the profit motive which would benefit a few vested interests. The use of land, technology, and taxes should be for the benefit of the people and not for the benefit of a few powerful interest groups. City-dwellers have to become politically active and organise themselves and agitate to change the existing economic and social systems in the cities.

Process of Urbanisation

Urban: What is an 'urban area' or a city or a town? This term is used in two senses – demographically and sociologically. In the former sense, emphasis is given to the size of population, density of population and nature of work of the majority of the adult males; while in the latter sense, the focus is on heterogeneity, impersonality, interdependence, and the quality of life.

The German sociologist, Tonnies (1957) differentiated between rural and urban communities in terms of social relationships and values. The rural gemeinschaft community is one in which social bonds are based on close personal ties of kinship and friendship, and the emphasis is on tradition, consensus and informality, while the urban gesellschaft society is one in which impersonal and secondary relationships predominate and the interaction of people is formal, contractual and dependent on the special function or service they perform. The emphasis on gesellschaft society is on utilitarian goals and competitive nature of social relationships. Other sociologists like Max Weber (1961:381) and George Simmel (1950) have stressed on dense living conditions, rapidity of change and impersonal interaction in urban settings. Louis Wirth (1938:8) has said that for sociological purposes a city may be defined as 'a relatively large, dense and permanent settlement of socially heterogeneous individuals'. Scholars like Ruth Glass (1956) have defined city in terms of factors like: size of population, density of population, main economic system, type of administration, and some social characteristics.

In India, the census definition of 'town' remained more or less the same for the period 1901-51; but in 1961, a new definition was adopted. Up to 1951, 'town' included (1) collection of houses permanently inhabited by not less than 5,000 persons, (2) every municipality/ corporation/notified area of whatever size, and (3) all civil lines not included within the municipal units. Thus, the primary focus in the definition of town was more on the administrative set-up rather than the size of the population. In 1961, certain tests were applied for defining a place as 'town'. These were: a) a minimum population of 5,000, b) a density of not less than 1,000 persons per square mile, c) three-fourths of its working population should be engaged in non-agricultural activities, and d) the place should have a few characteristics and civic amenities like transport and communication, banks, schools, markets, recreation centres, hospitals, electricity, and newspapers, etc. As a result of this change in the definition, 812 areas (with 44 lakh people) declared as towns in the 1951 census were not so considered in the 1961 census. The 1961 basis was adopted in the 1971, 1981 and 1991 censuses too for defining towns. Now demographically, areas with population between 5,000 and 20,000 are considered as small towns, those with population between 20,000 and 50,000 are considered as large towns, those with population between 50,000 and one lakh are considered as big cities, those with more than 10 lakh people are considered as metropolitan areas, and those with more than 50 lakh people are called mega cities.

Sociologists do not attach much importance to the size of population in the definition of city because the minimum population standards vary greatly. For example, in the Netherlands a minimum population of 20,000 is required for a place to be designated as urban; in France, Austria and West Germany, it is 2,000; in Japan it is 3,000; in the USA it is 3.500; and so on. As such, they give more importance to characteristics other than the population size. Theodorson (1969:451) has defined 'urban community' as "a community with a high population density, a predominance of non-agricultural occupations, a high degree of specialisation resulting in a complex division of labour, and a formalised system of local government. It is also characterised by a prevalence of impersonal secondary relations and dependence on formal social controls." According to Robert Redfield (American Journal of Sociology, January 1942), 'urban society' is characterised by a large heterogeneous population, close contact with other societies (through trade, communication, etc.), a complex division of labour, a prevalence of secular over sacred concerns, and the desire to organise behaviour rationally towards given goals, as opposed to following traditional standards and norms.

Urbanisation: Urbanisation is the movement of population from rural to urban areas and the resulting increasing proportion of a population that resides in urban rather than rural places. Thompson Warren (Encyclopaedia of Social Sciences) has defined it as "the movement of people from communities concerned chiefly or solely with agriculture to other communities, generally larger whose activities are primarily centred in government, trade, manufacture, or allied interests". According to Anderson (1953:11), urbanisation is not a one-way process but it is a two-way process. It involves not only movement from villages to cities and change from agricultural occupation to business, trade, service and profession, but it also involves change in the migrants' attitudes, beliefs, values and behaviour patterns. He has given five characteristics of urbanisation: money economy, civil administration, cultural changes, written records and innovations.

Urbanism: Urbanism is a way of life. It reflects an organisation of society in terms of a complex division of labour, high levels of technology, high mobility, interdependence of its members in fulfilling economic functions and impersonality in social relations (Theodorson, 1969: 453).

The Characteristics: Louis Wirth (1938:49) has given four characteristics of urbanism:

Transiency: an urban inhabitant's relations with others last only for a short time; he tends to forget his old acquaintances and develop relations with new people. Since he is not much attached to his neighbours, members of the social groups, he does not mind leaving their places.

Superficiality: An urban person has the limited number of persons with whom he interacts and his relations with them are impersonal and formal. People meet each other in highly segmental roles. They are dependent on more people for the satisfaction of their life needs.

Anonymity: Urbanites do not know each other intimately. Personal mutual acquaintanship between the inhabitants which ordinarily is found in a neighbourhood is lacking.

Individualism: People give more importance to their own vested interests.

Ruth Glass (1956:32) has given the following characteristics of urbanism: mobility, anonymity, individualism, impersonal relations, social differentiation, transience, and organic type of solidarity. Anderson (1953:2) has listed three characteristics of urbanism: adjustability, mobility, and diffusion. Marshal Clinard (1957) has talked of rapid social change, conflict between norms and values, increasing mobility of population, emphasis on material things and decline in intimate interpersonal communication as important characteristics of urbanism. K. Dewis (1953) has highlighted eight characteristics of the urban social system: social heterogeneity (people of different religions, languages, castes, and classes live in urban areas. There is also specialisation in occupation), secondary association, social mobility, individualism, spatial segregation, social tolerance, secondary control and voluntary associations. Lious Wirth (1938:124) has given five characteristics of urbanism: heterogeneity of population, specialisation of function, anonymity, impersonality and standardisation of life and behaviour. Though these characteristics present an exaggerated picture of the urban man and his life, an analysis there of is necessary.

Heterogeneity of Population: The large population in cities can be largely attributed to migration from different areas which leads to people of different backgrounds and beliefs living together. This mixture of people affects the working of informal controls – mores and institutions – and reliance on formally designed mechanisms for regulating the behaviour of individuals and groups increases. People no longer share the common sentiments, and being exposed to new ideas imported from other cultures through contacts with migrants, they challenge the out-moded beliefs and practices and adopt such new attitudes and lifestyles which help them in iinproving their economic status and coping with problems of adjustment. The influence of family and neighbourhood decreases and people come into conflict on the question of tightness of behaviour.

Specialisation of Function and Behaviour: The heterogeneity and the large size of population of a city favour the development of specialisation. Since the city has many facets of life and an individual can participate only in some of them, he becomes choosy and takes interest only in a few fields. Specialisation in function encourages a diversity of life patterns. Doctors, engineers, businessmen, lawyers, bureaucrats, factory workers, teachers, clerks, policemen, for example, have different life patterns, different interests, and also different philosophy of life. Each specialist group makes its own contribution to the community and thus a division of labour is created. The cloth merchant say, sells only cloth, and depends on many other specialists to manufacture, process and distribute cloth so that it reaches his shop. Such a division of labour permits an individual to benefit from a broader range of services than his own knowledge and capabilities provide. Each inhabitant in the city becomes dependent on specialists such as physicians, masons, mechanics, shopkeepers, tailors, washermen, and so on. He need not learn the techniques of each profession.

Specialisation provides to an individual diverse opportunities to act, to express himself and to develop his potentialities. However, the contacts become secondary and formal and the sense of living a common life and having common concerns is destroyed. The relationship between two persons remains for a short duration till they gratify each other's purpose.

In a social order characterised by a heterogeneous population and diversity in behaviour patterns, there is a greater likelihood of confusion among several alternatives for proper behaviour in a given situation. For example, a student finds other student using unfair means and getting first division. He then thinks, should he do the same thing? One person finds another person giving Rs. 10,000 and getting a job of a police sub-inspector. He becomes confused whether he should report the case and get the bribe-taker arrested, or should he develop an attitude of indifference? These moral, social, and legal dilemmas are overwhelming in the city life.

Anonymity and Impersonality: High population density in the city erodes a sense of personal identity leading to loneliness and a loss of a sense of belonging. Hundreds of people watch a movie in a picture-house, enjoy and laugh together but when the film ends, the common emotions disintegrate into anonymity and impersonality. On the other hand, this very anonymity is the crux of personal freedom. The lack of interest in others releases the individual from heavy pressures towards conformity. In many cases, his responsibility to others ends with payment. Even when he becomes a member of a voluntary group like a club for instance, his participation could be minimum. He does not have to win the acceptance of other members or to engage in the accommodative process of fitting himself into the frame of their expectations. He may observe others but he may not necessarily be carried away by their stimuli.

One advantage of anonymity is that individuals are not judged according to their parents' lower class status but are evaluated on the basis of their appearance and behaviour in casual contacts. The anonymity and impersonality of urban life gives an individual, who aspires for a higher status, a greater opportunity to take advantage of symbols of higher status, like wearing attractive clothing, improving his mode of speech and manner so as to gain the acceptance and to impress persons of high positions, maintaining contacts with them and ultimately achieving the goals he seeks through these contacts.

Standardisation of Behaviour: The urban life necessitates an individual to standardise his behaviour which ultimately helps him and others (with whom he interacts) to understand each other and make interaction simpler. For example, a shopkeeper finds the same questions being asked by a succession of customers. The customers are then seen as types — the person who haggles over price, the fellow who goes for quality, the man who is merely looking without any intention of buying, and so on. The experienced shopkeeper quickly judges the type of customer he is dealing with and uses the sales strategy he regards as most effective for a particular type of customer. This helps both the shopkeeper and the customer to handle the sales transaction in a simple and quick fashion. Such standardised expectations and behaviour are part of urban life. Markets, clubs, restaurants, buses, newspapers, TV, radio, and schools/colleges present a largely standardised picture. A person who is unable to fit into such a life finds himself out of step, and is faced with the problem of adjustment. The large size of the city population lends particular force to the standardisation of behaviour. This does not mean that the divergence of individual orientations is not possible.

Sorokin and Zimmerman (1962:56-57) have identified the following characteristics of the urban social system:

Non-agricultural Occupation: While agriculture is the main basis of the rural economy, trade, industry and commerce are the chief supports of the urban economy. It is this difference in occupation that ensures that rural people work in natural environments. Urban people on the other hand work mostly in artificial and unnatural environment in which the heat, cold and humidity are controlled by innovative skills. According to James Williams (1958), working in unnatural environment affects people's attitudes and behaviour patterns. It is, thus, because of the occupational differences that in urban areas we find liberals as well as conservatives, moderns as well as traditionalists, and unsociables as well as sociables.

Size of Population: Urban communities are much bigger in size than rural communities. The availability of job opportunities on one hand, and the materialistic as well as educational, medical and recreational facilities on the other hand attract people to cities.

Density of Population: In villages, people have to live near their fields to supervise the agricultural pursuits but in urban areas, people's residence depends on the location of their offices, market, children's school/college and so forth. This leads to a high density of population in areas which abound in these facilities. In India, the average density of population per square mile in metropolitan cities varies between 3,000 and 5,000 persons. This high density has its own benefits as well as disadvantages. The advantages are that social contacts multiply, all necessary facilities are easily approachable and selection of friends becomes easier. The disadvantages are that inhabitants have very formal and impersonal relations with each other and their mental stresses increase.

Environment: Bernard (1971) has talked of four types of environment: material (climate), biological (animals and plants), social, physiosocial (machines, gadgets, instruments) and psychosocial (customs, traditions, institutions, etc.) and composite (economic, political and educational systems). The urban environment is more polluted. Besides, because of being surrounded by educational institutions and hence being more educated, an urban dweller is more rational, secular, and competitive.

Social Differentiation: In urban areas, people are differentiated on the basis of occupations, religion, class, living standards and social beliefs. Yet, they are dependent on each other and act as a functioning whole.

Social Mobility: Urban areas provide opportunities for change in social status because of which, as compared to villages, there is more upward mobility in cities. The mobility may be horizontal or vertical. Besides social mobility, we also find geographical mobility in urban areas.

Social Interaction: Relations among urban inhabitants are secondary and impersonal. People are more concerned with the status and skills of other persons than with their beliefs and ideologies. Control is also so formal that many a time it creates deviant behaviour.

Social Solidarity: In comparison to mechanical solidarity in rural areas, there exists organic solidarity in urban areas. In such a solidarity, though each person has his own individuality and personality, he depends more on others for their specialised roles.

The above description of the characteristics of urbanism as a way of life gives a feeling as if personal relations, primary groups and social intimacy do not exist in the cities. If consciously developed organisations serve the interests of the individuals, the primary groups also admit members through birth. Primary-group members are tied together by a fusion of concerns for one another. Their relations are more emotional and intensive. Within the group, a member performs variety of functions unlike specified functions in secondary groups.

For example, in a family, a mother serves as a cook, a nurse, a moral instructor and a manager of tensions for the children and the family members. Although the social change has weakened the bonds of family, neighbourhood and peer groups yet the old kind of functioning of these groups has not completely stopped nor' have the primary relations vanished. Performing obligatory roles in the family, maintaining social participation within the neighbourhood, sharing common interests of castes and, acting as a source of support to one's kin and friends continue to be important and significant features of urban life. A number of studies in India (like those by Kapadia, Sachchidanand, R.K. Mukerjee and M.S. Gore) have shown that rural people who migrate to cities continue to maintain links with their families and kin in the village. In cities too, they not only share their problems with persons belonging to the same and adjoining villages but also with members of their caste. This makes their adaptation to city life easier.

The Growth: While cities have existed since ancient times, until recently they represented only a relatively small proportion of the population. The lives of the great majority of the people were predominantly shaped by the rural community or village. The massive growth of cities and metropolitan areas, and the shift of a significant proportion of the population to urban areas has been a characteristic feature of past six decades or so. Urbanisation was an offshoot of the industrial revolution which created a demand for a large number of workers at centralised locations.

The growth of cities depends not only on birth and death rates and migration but also on political, religious historical and economic factors. Political centres can be the capital of states (Bhopal, Jaipur, Bombay, Calcutta, etc.) or the areas of political activities (Delhi), or the training centres for the military (Kharagvasla), or centres for defence production (Jodhpur); economic centres are areas which predominate in trade or commerce (Ahmedabad, Surat); industrial towns are places with factories (Bhilai, Singrauli, Kota, Ludhiana); the religious cities are those where people go on pilgrimage (Hardwar, Varanasi, Allahabad); and educational centres have educational institutions (Pilani).

In India, the urban population in 1971 was 109.11 million, in 1981 it was 160.1 million, and in 1991 it was 217.18 million. It is expected to increase to 291 million by March 2001 (The Hindustan Times, January 19, 1997). While in 1921 the urban population was only 11.3 per cent of the total population of the country; in 1951 it increased to 17.6 per cent, in 1971 to 20.2 per cent, in 1981 to 23.8 per cent and in 1991 to 25.7 per cent (Census of India, 1991, Series 1, 2). Again, in the 1911-21 decade, the urban population increased by 8.3 per cent, in 1921-31 it increased by 19.1 per cent, in 1931-41 by 32 per cent, in 1941-51 by 41.4 per cent, in 1951-61 by 37 per cent, in 1961-71 by 38.2 per cent, in 1971-81 by 35.4 per cent, and in 1981-91 by 40.4 per cent. Thus, during the four and a half decades between 1941 and 1985, the growth rate of urban population was between 3.4 and 3.8 per cent every year. During

1985-90, the average annual growth rate for urban areas was 2.95 per cent as compared to 1.65 per cent for rural areas for the same period. The growth rate for urban areas will increase during the next two decades whereas the growth rate for rural areas will decrease. As per the UN projections, urban areas in India will be marked by a growth rate of 3.03 per cent during 2015-2020, while rural areas will have a reduced growth rate of 0.34 per cent during the corresponding period. Also as per UN projections, 45.2 per cent people (about 630 million) will live in urban areas in India by 2025 AD. This is close to the total population of India in 1981 which was 684 million (The Hindustan Times, November 29, 1995).

The share of agricultural employment in the total employment of main workers reduced from 72 per cent in 1961 to 64 per cent in 1991 (Census of India). By the year 2001, anything between 18 to 20 crore people would be added to the rural population and of this addition, at least 10 crore of them will come to urban areas in search of jobs.

Further, while in 1931, the total number of cities with more than one lakh population was 32, it increased to 107 in 1961, 216 in 1981 and 317 in 1991 (excluding Jammu and Kashmir). The number of bigger cities with more than ten lakh population increased from two in 1941 to nine in 1971,12 in 1981, and 23 in 1991 (Census of India, 1991, Series 1, Paper 2:251). Such growth affects the social, economic and political life of the people.

Of the total number of 4,689 urban settlements (in 1991) in India, Uttar Pradesh has the largest number of towns (704), followed by Tamil Nadu (434), Madhya Pradesh (327) and Maharashtra (307). According to the 1991 census, 65.2 per cent population was living in cities with more than one lakh population, 10.9 per cent in cities with population between 50,000 and one lakh, 13.2 per cent in cities with population between 20,000 and 50,000, 7.8 per cent in cities with population between 10,000 and 20,000, and 2.6 per cent in towns with population between 10,000 and 5,000 (Census of India, 1991, Series 1, Paper 2). Four major cities of India with more than 50 lakh population, according to 1991 figures are: Calcutta (109 lakh), Mumbai (126 lakh), Delhi (93 lakh), and Chennai (54 lakh) (The Hindustan Times, May 30,1991).

Effect of Society

The social effects of urbanisation may be analysed in relationship to family, caste, social status of women, and village life.

Urbanisation and Family: Urbanisation affects not only the family structure but also intra and inter-family relations, as well as the functions the family performs. Several empirical studies of urban families conducted by scholars like I.P. Desai, Kapadia, and Aileen Ross, have pointed out that urban joint family is being gradually replaced by nuclear family, the size of the family is shrinking, and kinship relationship is confined to two or three generations only. In his study of 423 families made in 1955-57 in Mahuva town in Gujarat, I.P. Desai (1964) found that 5 per cent families were nuclear (residentially as well as functionally), 74 per cent were residentially nuclear but functionally and/or substantially (in property) joint, and 21 per cent were joint in residence and functioning as well as in property. Of the 95 per cent joint families (joint in functioning and/or property, and/or residence), the degree of functionality was low in 27 per cent cases (that is, they were joint only in functioning), high in 17 per cent cases (that is, they were joint in functioning and property), higher in 30 per cent cases (that is, they were joint in functioning, property and residence but were two generation families), and highest in 21 per cent cases (that is, they were joint in residence, functioning and property and were three-generation families). This shows that though the structure of urban family is clanging, the spirit of individualism is not growing in the families.

Kapadia (1959) in his study of 1,162 families in rural and urban (Wavsari) areas in Gujarat in 1955 found that while in rural areas, for every two nuclear families there were three joint families; in urban areas, nuclear families were 10 per cent more than joint families. Ross (1961) in her study of 157 Hindu families belonging to middle and upper classes in Bangalore in 1957 found that:

1. about three-fifth (or 60%) of the families are nuclear and two-fifths (or 40%) are joint;
2. of the joint families, 70 per cent are small joint (couple+unmarried children+married sons without children, or two or more married brothers with children) and 30 per

cent are large joint families (parents of unmarried children unmarried children+married sons with children);

3. the trend today is towards a break with the traditional joint family form into the nuclear family unit;
4. small joint family is now the most typical form of family life in urban India;
5. there is a cycle of family types;
6. a growing number of people now spend at least part of their lives in single units; and
7. relations with one's distant kin are breaking or weakening.

R.K. Mukherjee (1973) also, on the basis of his study of 4,120 families in West Bengal in 1960-61, has said that replacement of joint family by nuclear family units is a fait accompli.

Though intrafamily and interfamily relations are changing, it does not mean that youngsters no longer respect their elders, or children completely ignore their obligations to their parents and siblings, or wives challenge the authority of their husbands. The important change is that the 'husband-dominant' family is being replaced by 'equalitarian' family where wife is given a share in decision-making processes. The parents also no longer impose their authority on the children nor do the children blindly obey the commands of their parents. The attitude of youngsters is motivated by respect rather than fear. I.P. Desai maintains that "in spite of strains between the younger and older generations,, the attachment of the children to their families is seldom weakened". M.S. Gore (1968) writes: "even in joint family, the eldest male consults his children and this consultation is not formal". Ross (1961), however, thinks that "the feelings of family obligation and emotional attachment to family members will almost certainly weaken and the authority of the patriarch break down. When this happens, there will be little left for identity within the larger kinship group", Our own feeling is that family in urban India (and for that matter in the whole country) will never disintegrate but it will remain a strong unit.

Urbanisation and Caste: Caste identity tends to diminish with urbanisation, education and the development of an orientation towards individual achievement and modern status symbols.

Urbanites participate in networks which include persons of several castes. According to Rajni Kothari, the structure of particularistic loyalties has been overlaid by a more sophisticated system of social and political participation with crosscutting allegiances. Andre Beteille (1966:209-10) has pointed out that among the westernised elite, class ties are much more important than caste ties.

The educated members of some castes with modern occupations sometimes organise as a pressure group. As such, a caste association competes as a corporate body with other pressure groups for political and economic resources. This type of organisation represents a new kind of solidarity. These competing units function more as social classes than as caste structures.

Yet other change we find today is the fusion of sub-castes and fusion of castes. Kolenda (1984:150-51) has identified three kinds of fusion:

1. on the job and in newer neighbourhoods in the city, persons of different sub-castes and of different castes meet. They are usually of approximately equal rank. Neighbourhood or office group solidarity develops. This has been found common in the government colonies in big cities;
2. inter-subcaste marriages take place, promoting a fusion of subcastes. This is because many a time it is difficult to find a sufficiently educated groom for an educated daughter within his own sub-caste, but one may find it in neighbouring sub-caste; and
3. democratic politics fosters the fusion of sub-castes and of adjacent castes. One example is the Dravida Munnetra Kazagam (DMK) and the Anna Dravida Munnetra Kazagam (ADMK) parties of Tamil Nadu composed of the members of higher non-Brahmin castes.

Urban dwellers do not conform to caste norms strictly. There is a change in commensal relations, marital relations, social relations, as well as in occupational relations. A study of caste system in Bihar revealed that urbanisation does did affect all characteristics of the caste system uniformly. On the basis of the study of 200 persons belonging to five different castes, it was found that all respondents had married in their own castes, though 20 per cent of the respondents living in cities (against 5 per cent in rural areas) were in favour of intercaste marriages. As regards occupation, not a single respondent in the city was engaged in his traditional caste occupation, though 81 per cent respondents in rural areas were still engaged in their traditional occupations. Likewise, caste solidarity was not as strong in urban areas as in the rural areas. Caste panchayats were very weak in cities. G.S. Ghurye (1952), K.M. Kapadia (1959), A.P. Barnabas, Yogendra Singh, R.K. Mukerjee, M.N. Srinivas, Yogesh Atal and S.C. Dube have also referred to the impact of urbanisation on caste.

Status of Women: The status of women in urban areas is higher than that of rural women. Urban women are comparatively educated and liberal. Against 25.1 per cent literate women in rural areas, there are 54 per cent literate women in urban areas according to the census of 1991. Some of them are working too. As such, they are not only aware of their economic, social and political rights but they even use these rights to save themselves from being humiliated and exploited. The average age of girls at marriage in cities is also higher than the corresponding age in villages.

However, in the labour market, women are still in a disadvantaged situation. The labour market discriminates against women and is opposed to equality of opportunity – understood in a comprehensive sense to include equality of employment, training and promotional opportunities. In this sense, change is not possible in the sex segregated labour market whose structures ensure that the career patterns of women will normally be marked by discontinuity, unlike the normal male career patterns which assume continuity. Because of the constraints of the sex segregated labour market, women tend to cluster in a limited range of occupations, which have low status and are poorly paid. Women normally prefer teaching, nursing, social work, secretarial and clerical jobs – all of which have low status and low remuneration. Even those women who have surmounted the hurdle to professional education are disadvantaged as they find it difficult to reconcile to the competing demands of a professional career and home.

It is difficult for women to remain single or to combine marriage with career. Apart from the general expectation that all wives must be housewives, it has been noted that women are called upon to sacrifice their career when the need arises, thereby subordinating their own career to that of their husbands. This often creates frustration among women, leading to psychotic illness in a few cases. Rural women, however, do not have to face such problems.

It has been further found that in the cities of India, the high level education among girls is significantly associated with the smaller family size. Though education of women has raised the age of marriage and lowered the birth rate, it has not brought about any radical change in the traditional pattern of arranged marriages with dowry. Margaret Cormack (1961:109) found in her study of 500 university students that girls were ready to go to college and mix with boys but they wanted their parents to arrange their marriage. Women want new opportunities but demand old securities as well. They enjoy their newly found freedom but at the same time wish to carry on with old values.

Divorce and remarriage are the new phenomena we find among urban women. Today women take more initiative to break their marriages legally if they find adjustment after marriage impossible. Surprisingly, a large number of divorces are sought by women on the grounds of incompatibility and mental torture.

Politically also, urban women are more active today. The number of women contesting elections has increased at every level. They hold important political positions and also possess independent political ideologies. It may, thus, be concluded that while rural women continue to be dependent on men both economically and socially, urban women are comparatively independent and enjoy greater freedom.

Urbanisation and Village Life: During the past half a century, urban development in our country has led to the centrifugal movement of village people to the urban areas that were located within fairly easy access of public utilities. Many migrated to cities because of the availability of jobs there. Those who continue to live in villages also enjoy many of the conveniences of city life, although they are miles removed from the urban centres. The excellent highways, automobiles, radio, television and newspapers keep the villagers in contact with the city culture and civilization. The combination of rural residence and urban employment, and urban residence and rural contact has resulted not only in certain modifications of social patterns but also in adjustments to a new way of life.

The villagers are now more aware of the city lifestyle and they have been influenced by it in such a way that they no longer lay undue emphasis on caste, creed... They have become more liberal in their approach. They no longer live in isolation. Many cultivators have accepted the new farm practices. Not only have their values and aspirations changed but there is a change in their behaviour too. The *jajmani* system is weakening and intercaste and interclass relations are changing. There is a change even in institutions of marriage, family and caste panchayats. Instead of depending on traditional methods of treating the diseases, they now use modern allopathic medicine. Similarly in elections, they give importance not as much to the social status of the candidate as to his individual potentialities and his political background.

But this in no way mean that traditions have ceased to be important in villages. Individualism as not been able to replace familism, nor has secularism been able to replace the bond with the sacred.

The Problems: Urban problems are endless. Drug addiction, pollution, crime, juvenile delinquency, begging, alcoholism, corruption, and unemployment are a few of them. Let us analyse the incidence and prevalence of seven crucial problems that are not covered in other chapters of this book. These are:

1. housing and slums,
2. crowding and depersonalisation,
3. water supply and drainage,
4. transportation and traffic,
5. power shortage,
6. sanitation and
7. pollution.

Housing and Slums: Housing people in the city 'houselessness' is a serious problem. Government, industrialists, capitalists, entrepreneurs, developers, contractors, and landlords have been unable to keep pace with the housing needs of the poor and the middle class people. According to the 1988 UNI report (The Hindustan Times, 9 May, 1988), between one-fourth and half of the urban population in India's largest cities lives in makeshift shelters and slums. At least 15 per cent of the nation's families are deprived of houses, more than 60 per cent of the houses have inadequate lighting and air facilities, and 80 per cent of the rural and 30 per cent of the urban population lives in mud-houses. Millions of people are required to pay excessive rent, that is, one which is beyond their means. In our profit-oriented economy, private developers and colonisers find little profit in building houses in cities for the poor and the lower middle-class people, and they consider it gainful to concentrate instead in meeting the housing needs of the rich and the upper-middle class.

The result has been higher rents and a scramble for the few available houses. Almost half of the population are either ill-housed or pay more than 20 per cent of their income on rent. In some states, the Housing Boards and the City Development Authorities have tried to remedy the city housing problem with active financial support from the Life Insurance Corporation, HUDCO and such other agencies. They even charge the total housing cost in monthly instalments on an interest varying between 9 per cent and 11 per cent. But engineers and contractors profit a lot from these government efforts. They use poor-quality material in construction and finish the houses contravening the laid down specifications. The buyer soon finds that the roof leaks, the plaster peels off, there are cracks in the walls, and the electric fittings break down. Such ventures bring bad name to housing boards and even of a few honest bureaucrats associated with such housing schemes. No wonder, housing in the cities even today continues to be a gigantic problem next only to food and clothing.

The estimated shortage of houses at the beginning of the Eighth Plan was about 30 million units, out of which about 8 million were required for the urban areas. By 1997, the shortage was expected to grow to 12 million units in urban areas. In Delhi alone, which has seen a population increase from 6.2 to 9.3 million between 1981 and 1991, there is an addition of 60,000 people each year who need to be provided with new housing. Almost 70 per cent of Delhi's population, according to an UNI report, lives in sub-standard conditions. With the country's slum population of 1991 standing at nearly 40 million, slum dwellers form 44 per cent of the population in Delhi, 45 per cent in Mumbai, 42 per cent in Calcutta, and 39 per cent in Chennai. The situation is no better in the eight other metropolises of Bangalore, Hyderabad, Ahmedabad, Kanpur, Pune, Nagpur, Lucknow and Jaipur. The slum population, governmental efforts notwithstanding, is expected to show a sizable increase by 2000 aggravating to the housing problem and the squalor conditions. The living conditions in slum areas are characterised by overcrowding, poor environmental conditions, scarcity of health and family welfare services and total absence of minimum level of residential accommodation. As a result, conditions of people living in slums is far more pathetic than in rural areas.

Let us take a look at the slum-dwellers of Bangalore, for an example. According to a survey conducted here in 1992, this Garden City has 464 slums with a total slum population of 8.51 lakh. Only 10 per cent of the slum population here has the facilities for sewage disposal. About 55 per cent of the slum habitats live below the poverty line with a monthly income of less than Rs 1,000. The average age of marriage is 18.3 years, as compared to 20.3 years for urban areas of Karnataka as a whole. The urban areas of Karnataka have an infant mortality rate of 42 per thousand but the slum dwellers of Bangalore have infant mortality rate of 77.8. Similarly, crude birth rate is 32.5 against 23.1 for urban Karnataka as a whole. Nine out often pregnant women are in the age group of 15-29. The concept of spacing births has not gained acceptance. Only 40 per cent children between the ages of one and two are fully immunised. The incidence of diseases is high among children. On an average a slum household spends 5 per cent of its income on medicines, private doctors fees and transportation for treatment.

The reasons for not going to government municipal hospitals are distance, long waiting time and rude behaviour of staff (The

Hindustan Times, November 29, 1995). This is the scenario obtaining in the slums of Bangalore, despite the fact that the Government has provided a network of 92 dispensaries, 16 hospitals, 3 pediatric centres, 37 Urban Family Welfare Centres, 31 maternity homes, two maternity hospitals, and 200 Anganwadis. We only need to strain our imagination to visualise the plight of slum-dwellers in other cities. The order of development in squatter settlements is people, land (sites), shelter and services. People first select a site which meets their social and economic needs, build shelters and then wait for the services to move in over a period of time. Although the settlements fulfil the needs of people, they violate city planning regulations. It is, therefore, believed that the current order of development ought to be land (site), people, shelter and services. Now the government, apart from encouraging the poor for going in for low cost non-formal housing technologies, has formulated several plans and given many concessions to promote more and better housing. This includes contribution of Rs 100 crore to the National Housing Bank, setting up a separate Social Security Fund with a corpus of Rs 100 crore, and creating a National Scheduled Caste and Scheduled Tribes Finance and Development Corporation.

Crowding and Depersonalisation: Crowding (density of population) and people's apathy to other persons' problems (including their neighbours' problems) is another problem growing out of city life. Some homes are so overcrowded that five to six persons live in one room. Some city neighbourhoods are extremely overcrowded. Overcrowding has very deleterious effects. It encourages deviant behaviour, spreads diseases, and creates conditions for mental illness, alcoholism, and riots. One effect of dense urban living is people's apathy and indifference. City dwellers do not want to 'get involved' in other people's affairs. Some people do take interest in accidents and in cases of molestations, assaults and even murders but others choose to be mere onlookers.

Water Supply and Drainage: We have reached a stage where no city has round the clock water supply. Intermittent supply results in a vacuum being created in empty water lines which often suck in pollutants through leaking joints. Cities like Chennai,

Hyderabad, Rajkot, Ajmer, and Udaipur get water from the municipality for less than an hour a day. Many small towns have no main water supply at all and are dependent on tube wells. Even a relatively planned and serviced city like Delhi has now to reach as far as 180 km to the Ramganga for augmentation of water supply. Bangalore pumps water from far away with a lift of about 700 metres. Most towns and cities which normally get good rain every year, have been undergoing the agony of acute water shortage in the last eight-nine years. What seems to be sadly lacking is a national water policy which would assess the total water resources and then allocate water. This is in spite of the State Chief Ministers' meeting at Delhi in September 1987 which approved the National Water Policy which aimed at giving priority to drinking water requirements.

When we look on the other side of the water supply, that is, drainage, we find the situation equally bad. One of the less known facts about India is that there is not a single city which is fully sewered. Not even Chandigarh can claim this distinction because the unauthorised constructions in and around it lie outside the purview of the main system. Because of the non-existence of a drainage system, large pools of stagnant water can be seen in every city even in summer months. Just as we need a national water policy, we also need a national and regional drainage policy.

Transportation and Traffic: The transportation and traffic picture in all Indian cities is unhappy. A majority of people use buses and tempos, while a few use rail as transit system. The increasing number of scooters, motorcycles, mopeds and cars make the traffic problem worse. For example, in Mumbai, automobiles have trebled (from 3.1 lakh to 8.73 lakh) and in Delhi increased by four times (from 5.92 lakh to 23.7 lakh) between 1986 and 1996 (The Hindustan Times, November 29, 1996). They pollute the air with smoke. In Mumbai alone, the daily pollutants let out into the air is about 3,000 tons, of which 52 per cent come from automobiles, 2 per cent from domestic fuels, and the remaining 46 per cent from industries.

The number of buses plying in metropolitan cities like Delhi, Mumbai, Chennai and Calcutta is not adequate and commuters have to spend about one to two hours to get into a bus, which means leaving the house two hours in advance in the morning to reach their place of work and reaching home two hours late in the evening. The main reason for being in this mess is that the low income of the commuters forces them to live in areas with cheap accommodation which necessitates extensive travel. Further, since our citizens cannot afford to pay high fares for the use of a public transport system, the fares have to be kept very low which results in all city bus services sustaining such annual losses as hamper their expansion or maintenance a fleet adequate to meet city needs.

Power Shortage: Closely linked with transportation is the question of power shortage. The use of electrical gadgets has increased very much in cities, and on the other hand, the establishment of new industries and the expansion of the old industries has also increased dependence on electricity. Most states are not in a position to generate the power that they need with the result that they remain dependent on the neighbouring states. Conflict over supplying of power between two states often creates severe power crisis for people in the city.

Sanitation: Municipalities and municipal corporations in Indian cities are so riddled with corruption and inefficiency that they have time for all things but the interest in sanitation of the city, particularly in removing garbage, cleaning drains, and unclogging sewers. The sweepers rarely and reluctantly perform their assigned duties and every few months threaten to go on strike on the issue of wages, etc. The garbage disposal fleets operate to a third or half of their capacity. If removing garbage work is assigned to private contractors, they always complain of non-payment of money and stop working on slight pretexts.

There is, thus, total lack of motivation to tackle the basic sanitation needs of the cities. The spread of unauthorised slums in congested urban areas and lack of civic sense among the settlers in these slums further adds to the growing mound of filth and diseases. Even the major cities of our country reflect the sanitation crisis. Of about 3,000 to 5,000 tonnes of garbage generated in one day in a metropolitan city, hardly 50 to 60 per cent is cleared. Out of 1,500-2,000 million litres of sewage generated in a day, hardly 1,000 to 1,500 litres a day is collected. This is when the total budget of municipal corporations of cities like Delhi, Calcutta, Mumbai and Chennai varies between Rs 1,000 and Rs 1,500 crore per annum (India Today, October 31, 1994:63-66). Municipal machinery works only in the VIP areas, totally neglecting the sewage, water or garbage collection in other colonies. According to UNICEF, lakhs of urban children die or suffer from diarrhoea, diphtheria, tetanus, and measles, etc. because of poor sanitary conditions and water contamination.

Many cities report a good number of malaria and filaria cases every year. More than half our battle against these diseases can be won if our municipal officials sincerely devote time to cope with sanitation problem and if our people become more responsible to the need for cleanliness. The outbreak of the plague in some cities in our country in 1994 has created some awareness among the officials and the general public but covering the whole city by sewage, arranging the disposal of garbage, cleaning drains and community toilets, etc. are such big problems that clean environment, community hygiene and city sanitation still remain and are likely to remain serious problems in decades to come.

Various forms of racketeering in municipal work exit in our cities. For instance:

1. since payment for garbage removal is made on the basis of trips and not the weight of the garbage pickup vehicles, a large number of trips are shown on records and money is split between the contractor and the municipal employees;
2. a large number of vehicles used for garbage-collecting operations are actually used for outside work;
3. debris is diverted and sold to private parties for filling up building sites while payment for debris disposal is also taken from the municipality; and
4. drivers of trucks and dumpers sell diesel meant for sanitation trucks.

Obviously, the basic problem is excessive urbanisation and the resultant slumming of the cities. But because our politicians use migrants as vote-banks, they remain unconcerned with taking necessary civic action. Lack of understanding at the planning level, lack of coordination between the concerned agencies, mismanagement of municipalities, and lack of necessary funds to be provided by the state governments will always remain barriers in putting sanitation maintenance cycle in proper shape. If cities continue to treat sanitation and sewage as low-priority areas, over coning health crisis in urban areas will be an insurmountable task in years to come. As a long-term remedy, what is needed is using new techniques of refuse collection, new technology for the disposal of garbage, and a fundamental change in the municipal infrastructure, and land use planning.

Pollution: Our cities and towns are major polluters of the environment. Several cities discharge 40 per cent to 60 per cent of their entire sewage and industrial effluents untreated into the nearby rivers. The smallest town contributes its share of garbage and excreta to the nearest waterway through its open drains. Urban industry pollutes the atmosphere with smoke and toxic gases from its chimneys. Areas recording higher levels of air pollution abound with many ailments which particularly affect children below five years and people above fifty years of age. The high synergistic effect of sulphur dioxide, nitrogen dioxide, etc. causes these diseases. The ambient air quality in Delhi gives it the dubious distinction of being the fourth most polluted city in the world.

The issue of environmental pollution in urban areas is considered so significant that even the Supreme Court ordered in July 1995 for the strict enforcement of environmental laws leading to closure or relocation of about 146 hazardous industries in Delhi by November-December 1996. The orders led to even agitations among the affected workers in December 1996, but the Apex Court stuck to its decision of not locating industries within the National Capital Region (NCR) but shifting them by where in the neighbouring states. The vehicular emissions in Delhi account for 64 per cent of Delhi's air pollutant load, power plants are responsible for 16 per cent and industries account for 12 per cent. The closure of industries in Delhi was however stayed until June 1997.

Similar significant role was played by the Supreme Court in yet another judgement on environment protection in Andhra Pradesh in October 1996. The judgement banned shrimp (small marine fish) culture within 500 metres of the high tide line along the 6,000 km coastline and dismantling of all structures within the restricted zone by March 1997. The country earns more than Rs 600 crore in foreign exchange through export of shrimp from Andhra Pradesh and Tamil Nadu alone. According to an estimate by the Central Pollution Control Board, the effluent generation from aquaculture farms in the east coast alone was about 2.37 million cubic metres (The Hindustan Times, January 6, 1997). The poison we put in the environment comes right back to us through our air, water and food, slowly seeping into our bodies and showing up as cancer, immune disorders or as hormonal system disorders. No wonder, the doctors claim that the worsening environmental condition in India has increased the chances of catching cancer dramatically in the four metros – Delhi, Mumbai, Bangalore and Chennai – the chances of having cancer are as high as 7 to 11 per cent during a life time. A conservative estimate of cancer patients in India by the year 2001 stands at eight lakhs. Of course, the poor suffer more than the rich from environmental degradation.

The Causes: Following McVeigh and Shostak (1978:198-205), who have linked urban problems in the United States to four factors, we can identify following five major causes of problems of urban life in India:

1. migration in and out of the city,
2. industrial growth,
3. apathy of the government,
4. defective own planning, and
5. vested interest forces.

Migration: As already indicated, people migrate to towns/ cities because of the relatively better employment opportunities available there. In India, the migration has four patterns: rural to rural, rural to urban, urban to urban and urban to rural. Though rural to rural migration is by far the most prevalent form of movement but rural to urban and urban to urban migration is equally crucial. The 1971 census figures pointed out that in 71.3 per cent cases, migration was from rural to rural, in 15 per cent cases, it was from rural to urban, in 8.8 per cent of cases it was from urban to urban, and in 4.9 per cent cases, it was from urban to rural (Bose, 1979:560). The figures must be approximately the same today (in 1997). The analysis of intradistrict migration (short-distance migration), interdistrict or intrastate migration (medium distance migration) and inter-state migration (long distance migration) shows that about 68 per cent migrations are short distance, 21 per cent are medium distance and 11 per cent are long distance migrations (Bose, 1979:187).

The entrance of the rural poor into the city depletes sources of revenue. On the other hand, the rich people today prefer to live in suburban areas. This movement of the rich causes financial loss to the city. This migration to the city and away from the city aggravates problems.

Industrial Growth: While the urban population growth rate is 4 per cent in India, the industrial growth rate is about 6 per cent per annum. The Eighth Five Year Plan postulated an industrial growth rate of 8 per cent per annum. This growth was expected to take care of the additional job requirements in the cities. The tertiary sector also provides refuge to the migrants, though their earnings remain at low level.

Apathy of the Government: The administrative mismanagement of our cities is also responsible for the mess in which city-dwellers find themselves. Municipal governments have not kept pace with city growth, either spatially or in terms of management infrastructure. There is neither the will nor the capacity to plan for the future. There is also no skill and capability to manage what exists. Until we improve the capacity of our cities to govern themselves, we cannot emerge from the urban mess. On the other hand, the state governments also put many restrictions on local governments in raising necessary funds for dealing with particular urban problems.

Defective Town Planning: A more alarming factor in the general deterioration in the standard of civic services is the growing sense of helplessness of our planners and administrators. From the Planning Commission downwards, there seems to be a fatalistic acceptance of the uncontrolled growth in our metropolitan cities. In fact, a member of the National Commission on Urbanisation had expressed a feeling that very little is being done in our country to plan the growth of the cities in a proper way.

Vested-interest Forces: The last cause of urban problems is the vested interest forces that work Against people but enhance private commercial interests and profits. The city residents are usually powerless to affect decisions that the elite take to further their own interests, power and profit. When these powerful elite can make more money, they adopt plans and programmes no matter how many people are hurt in the process.

The Solutions: Some measures have to be adopted if we want to remedy urban problems.

Womens' Education

In today's World, a modern educated women suffers from various tensions. If she is a housewife she feels stressed because her education has not incurred her any relief from domestic chores. If she is working woman she is stressed because of lack of coordination between her home and office life. This chapter identifies stresses typical to the modern educated woman and offers some suggestions for dealing with them.

The ambivalence that women feel as they take on dual roles of housewife and worker is quite tension producing. Undoubtedly, many women are able to combine both roles with smooth adjustments, particularly if the husband and the family members are cooperative and supportive. However, in many families, the conflict between both the roles has created estrangement between either the husband or the wife or the wife and the other members of the family.

We have traced the psychology of the Indian women in terms of her desire for domination and her will for subordination. The large number of women who have traditional outlook are meeting with their tensions and stresses by overtly subordinating themselves to men and covertly trying to dominate them by their extreme devotion and sense of duty. The modern emancipated women are revolting against this duplicity. They are denying the superiority of the male and seeking an equality between the sexes overtly and covertly both. We have also advocated that through our educational system the women should learn to move towards equality of sexes without involving themselves in any duplicity in their behaviour pattern. But the danger in such a move is the enhancement of tensions and stresses in the individuals, in the families and also in the society.

Ambitious men of lower income group want their wives and daughters to work and earn to help raise the family's standard of living. At the same time these men cannot tolerate the idea of independence such as that they themselves have on the job. They are unable to free themselves from the myths of Sita and Savitri. They become agitated when their women become assertive and begin to express their personal views and opinions and their likes and dislikes. They feel that women are transgressing their freedom, which they have so generously bestowed on them. Their ego is hurt if their wife's income exceeds their own Income. But the women when they go out for work cannot be bound in the chains of orthodoxy. Hence the modern woman's tensions enhance as she has to cope with the demands of her job and the dictates of her male relations.

In a study by Mukta Mittal on "Educated Women Power," it was found that the need for supplementing the family income was the chief motivating factor for encouraging the respondents belonging to "lowly educated" and "moderately educated" groups to become job seekers and get themselves registered at the Employment Exchange Bureaus. But in the case of "highly educated" respondents, the prime motivating factor behind encouraging them in their becoming registrants for job was "desire to be free from dependence on family members and relatives".

The Perceptiveness

The Indian women are marching ahead in each and every field of work and activity. They are in the forefront of all the movements for progress and development. But unfortunately the area of social reforms, which touches them directly, is as yet a neglected area. The movement for women emancipation and empowerment are ridiculed and made fun of by the traditionally oriented males. The women have little say in the Parliament or state legislatures because their number is extremely limited in these bodies. The Bill for 30 per cent reservation of seats for women in the Parliament is opposed on one flimsy ground or another. Our Parliament and legislatures are responsible for the governance and the administration of the country. On them are needed such persons who are highly motivated towards efficient management, adequately educated, well-informed and possess a zeal for social service. The country at present has quite a substantial number of such women who can adorn the seats of the Parliament. There are, however, some leaders who want those women in the Parliament who are mere appendages of their husbands or male relatives. The ridiculous thing is that they want to do it in the name of social justice.

The stresses and the tensions of the modern educated Indian women can be minimised if they involve themselves in a strong movement of women emancipation and empowerment. In the Indian context it means the deliverance from the myths which have been woven around them and the false pedestal at which they have been put since long.

In 1963 Frieden wrote the Feminine Mystique which attempted to explode many myths about women. Her thesis is applicable to both East and West. She writes- 'the Victorian culture did not permit women to accept or gratify their basic sexual needs, our culture does not permit women to accept or gratify their basic need to grow and fulfil their potentialities as human beings, a need which is not solely defined by their sexual role". Just as the Indian women have been led to believe that their fulfilment lies in motherhood and wifely duties, so Friedan observes about American women: "In the feminine mystique, there is no other way she can even dream about herself, except as her children's mother, her husband's wife."

Simone de Beauvoir considers that the myth of feminine 'mystery' has several advantages for the male. He can dismiss inexplicable moods, behaviours, and feelings by saying:" Oh Women! Who can understand them any way?" Rohrbaugh Joanna adds: "Since woman cannot be understood, man cannot be expected to build an authentic relationship with her. He is free to relate to her in terms of his own perceptions, fantasies and desires".

Williams referring to the myth of woman as mystery says: "By defining her as mysterious other, man spares himself the necessity of analysing her behaviours and understanding it as a consequence of her position vis-a-vis him. To do that would require acknowledgement of her oppression, and a possible shift in their power relationship? The price would be very high".

In such a 'mystique' or lifestyle, a woman's intelligence has no meaning and her capability is negated. By following a monotonous daily routine, she loses all interest in the outside world, draws herself into a shell and spends her time in gossips. She is little more than an instrument, which keeps a constant supply of members through her powers of reproduction.

Friedan emphasises that one's individuality when becoming a wife and mother should not be given up, but strengthened. She says: "Maslow found that the individuality is strengthened, that the ego is in one sense merged with another, but yet in another sense remains separate, and strong as always. The two, tendencies, to transcend individuality and to sharpen and strengthen it, must be seen as partners and not as contradictory". Self-actualised educated women should serve their society and contribute their best to humanity, just as men have the opportunity to do so. The world has suffered enough because half of humanity has remained dependent on the other half.

Indian women are often viewed as passive, prudish and fragile. Many of them feel that a mere lustful glance or touch by a non-relative male leads to the loss of their chastity. The women's movements should actively try to alter such a nonsensical perception. It may be remembered that the women in general and the educated women in particular are in danger of becoming not only household slaves but also work place slaves if they do not learn to become individualistic and assertive.

An Indian woman with lustre with a brilliance of her won, with a dynamic attitude towards life and with the charm of a sturdy woman going about her business is always in danger of being branded as aggressive, too outgoing and, perhaps, flirtatious, This type of woman is incongruent with the image of the ideal that has been imprinted on the minds of the Indians for the centuries.

The women's movements should be in the direction of relieving intelligent, modern women from the clutches of orthodoxy. The real women's movement which India need and towards which the Indian feminists seem to be fully conscious is that of removal of orthodoxy, blind faith, poverty, helplessness, dependence and complete subjugation of woman's will. The reformers should try to bring the changes in the attitudes of both male and female regarding marriage and the family. The must attack fanatical religious rites, ill conceived notions of virginity, wrong notions of chastity, custom of giving or taking dowry, wastage in the marriage ceremonies and similar other outdated customs and traditions.

We may agree with Carden when he says that: "The new feminism is not about the elimination of differences between the sexes, nor even simply the achievement of equal opportunity; it concerns with the individual's right to find out the kind of person he or she is."

Decisive Conditions

A very serious situation is developing in some places of work where the boss is male and he tries to sexually exploit his female employees. The women organisations, the government and even the courts are seized with this problem. Some women take the courage to make complaints about this behaviour of their officers or male colleague but there are lots of women who silently suffer at the hands of office sharks. Sometimes men also become victims of women's manipulations. They make false complaints about their exploitation so that the persons concerned are defamed. It is because of this danger that the complaints of women regarding exploitation are being scrutinised cautiously. Such situations can be avoided in case men and women both are made to realise that there shall be no discrimination in the work situations on the basis of sex. Both men and women must view each other as equal partners in work performance and must expect that recognition shall depend on their efficiency and quality of work and not on their sex.

As has already been pointed out that greatest danger resulting from a woman being career oriented is disharmony within her family. In Kapur's study of much marital maladjustment, the husband expected the wife to work as well as serve him and the household. The husbands generally believed in absolute supremacy over their wives and desire complete surrender and devotion from them. Those women who asserted their individuality were severely punished and ruthlessly treated, In cases where women could not tolerate the brutalities, they were separated from their spouses.

If the husband demands that his wife quit work and she refuses, it creates much tension in the family. In those cases in which the husband is transferred and the wife refuses to quit her job and go with him, legal problems arise. The husband may sue his wife for the restoration of conjugal rights. In such cases the attitude of the Indian courts has not been very decisive. Kusum (1976) cited many cases in which the wife was asked to leave her job and join her husband. In one case (Gaya Prasad Vs Smt. Bhagwati) in Madhya Pradesh, the wife worked as Gram Sevika due to financial circumstances. The husband petitioned for the restitution of conjugal rights. The observations of the court while delivering the judgement were: "Merely on the ground that the husband has a small income, and the wife if allowed to serve at a place away from the marital home, can substantially augment the family, cannot be held to be a sufficient reason to deny the wife's society to the husband".

The above judgement was given in the sixties. From seventies onwards the change in the attitudes of the courts is noticeable, as judgements have become much less traditional. The judges concede that in case of genuine economic necessity, the wife has a right to maintain her job. But the main problem here is not only the economic necessity but also a woman's freedom to take her own decisions. The woman's personal satisfaction, sense of confidence and security and her intellectual needs must be the guiding factors in making a decision about her quitting the job or remaining in service while living away from her husband or family. Whatever may be the decisions of the courts the stressful life between the husband or and wife will continue to be lived till there is the recognition that wife also has aspirations, ambitions and a will of her own. The idea that the women should work only to supplement their husbands income is to give them much inferior position in the family than that of husbands and of considering them as only the instruments for augmenting the family incomes. The women should not be viewed as the objects or instruments, but as entities unto themselves.

Narrow and conservative concepts of in-laws are also tension producing. The in-laws, in many cases want that their daughters-in-law take up jobs but at the same time are extremely critical if she is late in returning from her place of work or is not able to do the household work as efficiently as they expect her to do.

Some studies indicate that women seek work equal to the level of their husband's prestige. In case a wife is not equally educated or trained she may prefer to stay at home rather than accept work below her husband's level. In such a case if she is forced to undertake inferior status work, she remains tense and suffers from inferiority complex. The woman herself can take steps to come out of such situation by undertaking courses or training to improve her qualifications.

There is another side of the picture also. If the wife is employed in a job which has higher prestige and emoluments than the husband's job, the husband feels jealous and threatened. In many cases the tension between the husband and wife becomes so intense that living under the same roof becomes a torture leading to divorce or separation or to perpetual quarrels between them. There can develop proper amity and understanding between them if the wife tries to understand the sentiments of the husband and the husband realises that his wife deserves what she is getting. Truly well adjusted couples are those who respect each others individuality. A husband and wife can achieve harmony through mutual support and self-esteem. When both work role differentiation should not exist. The husband should give full emotional support to the wife and the wife should be considerate to the needs of the husband, may they be either psychological, physical, sexual or economical.

The girls including those who are educated find it very difficult to get a husband who does not demand dowry. In fact now the eligible bachelors and their families are demanding that the girls be educated so that they can earn money and also a dowry to meet the initial expenses of setting the home. Dowry demands are increasing day by day with the rising ambitions for equipping the houses with the modern gadgets. The educated girls whose parents search bridegrooms who are well-employed and highly educated face demands of exorbitant amounts as dowries. Such demands many parents are not able to meet with. The sensitive girls feel themselves as the cause of their parent's woes and blame themselves for being born as girls. Sometimes their tensions increase so much that they even commit suicide.

The statistics relating to dowry are very grim. Sometime back it was announced in the Lok Sabha that as many as 878 cases of dowry murders and 1,479 cases of dowry suicides were registered in the country in 1990. The educated girls who resent their husband and in-laws demands of dowry are brutally killed or forced to commit suicide. In order to stop these great human tragedies and combat horrible social evils, it would be necessary to bring about radical transformation, in the old rigid social structure. The parents of the educated girls rear them up in such a rigid environment that they themselves are incapable of finding their own life partners. They have to depend on the traditional system of marriage. The educated girls may be saved from much stress if they are left free to make a choice of their own mates. The rigid social structure can be changed only when a brave new philosophy of life steeped in socio-economic and moral values of equality of all castes and both the sexes is evolved out and adopted. Such a philosophy will enable the male as well as the female child to spontaneously internalise the principle of equality of man, woman of all castes and creeds.

Contradicting Approach

Many conflicts arise when both husband and wife are working. The husband desires that the wife should take up job but disapproves the complete involvement of the wife in the job. For him the job of the wife is only a secondary commitment. Her primary duty he feels is towards him and his family. He wants his wife to work and also to take up full time duties of looking after the household. When the wife is unable to cope with this situation the seeds of conflict and estrangement between the husband and wife begin to germinate. After returning from her work the wife is tired but she receives no sympathy or help from her husband or in-laws. When the tensions so created become unbearable the wife has no option but to resign from the job. But this is also not liked by the husband who resents the loss of income and blames the wife for not being able to reconcile between her job outside and her duties to her family.

The tension reduction in such cases is possible when all the family members recognise that the household work is to be shared and is a joint responsibility of all the members of the family. In those families in which the joint responsibility is recognised the tensions are suitably dealt with.

Till nearly three decades earlier the women were giving utmost importance to the role of the housewife. In a study by Cora Vreede Stuer in 1970 it was reported that quite a substantial number of girls "consider the role of the housewife as the most suitable." But when they were asked that if they had to go outside the home to work, most of them said they would prefer to teach or do social work. One girl in Cora's study remarked: "I would be willing to work in my field, but if my husband opposed it, I would submit to his wishes." Similar views were expressed by 18 of 15 educated women in Rama Mehta's study of the Western Educated Hindu women in 1970. One woman in her study expressed this view: 'Working was not important enough for me to go against my husband's wishes. After marriage, one cannot do as one wants, working is not an important enough issue to create tensions". It may be noted that till seventies or may be later also the working women, generally, perceived their main role as that of wife or mother. By such an attitude they were meeting with their tensions created by their becoming working women. But the situation at the end of the twentieth century is not as simple as that. The aspirations and the motivations of the women have increased manifolds and so also there is increase in their stresses and tensions.

Society now has a more favourable attitude towards the employment of women in the middle income groups. It has become an economic need. Promilla Kapur in her study in seventies took a sample of 300 working women from three major occupations — teachers, office workers and doctors. She observed that because of society's change of attitude, as well as change in attitudes of the educated married women towards their own employment, their number has multiplied to the extent that they now constitute a class by themselves. This class is facing the greatest change and, challenge ever offered anywhere in the world to the feminine population.

The number of women seeking jobs has increased manifolds now from that of seventies. The challenges before them have also become quite serious. The educated women have first to face the spectre of unemployment. The job market is very tight. It is difficult for men to get the jobs and when women also compete with them the number of the job seekers becomes quite high. With the limited job opportunities the women also suffer from the various types of restrictions which are put on them by the society. They still search the jobs, which are considered feminine in nature like the teaching, or nursing or office work, etc. They also want a job in the town or place to which they belong. They do not opt for the jobs in the villages or remote areas. It is not only due to the intention of the women that they do not want to work at the distant places from their homes but also due to the living conditions. It is difficult for a single working woman to find a decent living place in most of the towns or big cities. The security environment in the country is also not such as the women may move very freely. The cases of eve teasing are on the increase and the single working woman becomes an easy prey of unscrupulous males.

The women's problems are threefold with regard to getting employment. First of all, they have to compete in an overcrowded job market. Secondly, their families and society do not allow them to serve in those professions, which are branded as predominantly masculine in nature, even though there may be no restriction from the side of the employing authorities. Thirdly, they have the problem of finding suitable living accommodation if they get a job away from their homes or native place and added to this is the problem of security for a single working woman. All these problems are stress producing in the educated women. But it may be said that the stresses of the women can be reduced if they themselves and the society approach towards the solution of the problems with some dedication. There is no doubt in it that the number of jobs has to be increased. More efforts should be made to educate the women to generate self-employment. They may also be given vocational training. A change is to be brought in their own attitudes and the attitudes of the society and the family that it is a myth that the women are incapable of taking up some jobs which are masculine in nature. The women are capable of taking up all those jobs, which are considered masculine. In previous pages we have forcefully built the case of sexual equality on the basis of the physiology of the male and the female. Lastly, the government and the society must come forward to build "Working Women's Homes." In some towns they have been built. The need is to have a chain of such homes in each town and township.

There is another type of conflict with which the working women suffer. This is in relation to their work environment. The working women rightly demand that they should be accepted and respected as equally capable and efficient workers as males. The conflict occurs when they simultaneously demand special privileges and advantages because they are the women and the weaker sex. For example, they take up jobs that require night duties but after taking the job they may claim that since they are ladies they may be exempted from working in the night. They may press the employer to transfer them to some daytime job and transfer some male to this job. This becomes a conflict-producing situation. Some women also shirk work claiming that they are women and so entitled for light work. To avoid such conflicts the women need to develop professional attitude. There are many examples of successful women.

There is another side of the picture as well. The men resent the working women in their midst. In an office in which men and women both work the men do not feel as free with the women as with their men colleagues. If male becomes too intimate gossips are unleashed. If he ignores them he is branded as the chauvinistic male. There seems to be a need to develop a code of ethics for interpersonal relationships in those situations in which the males and females have to work in close coordination.

10

Democratic Aspects

Abraham Lincoln defined democracy as a government of the people, by the people and for the people. Despite the frequent attempts at defining democracy, this is probably the most accurate description. In a democracy one finds a government of the people which is made up of the elected representatives of the people, elected on the basis of adult suffrage. The ideals of democracy are liberty, equality and fraternity. Democracy aims to establish political, economic and social equality, and gives every individual the constitutional right to express his own opinion, to associate with any group, to indulge in any legitimate action.

The final objective of democracy is not merely successful government but the creation of an ideal society in which people have the greatest chance of evolving brotherhood. Democracy seeks to create an environment which is conducive to the highest and most beneficial development of the human personality. The principles of democracy were described in the following words as part of America's Declaration of Independence in 1776: "We hold these to be self evident that all men are created equal, that they are endowed by their creator with certain inalienable rights, that among these are life, liberty, and the pursuit of happiness, that to secure these rights, governments are instituted, deriving their just powers from the consent of the governed." It was implied that men were born equal in respect of their rights and that they would remain so. The objective of political society is to defend man's natural and unobtrusive rights; the right to freedom, property, security and opposition to injustice. The element of absolute authority basically inheres in the nation. No organisation, no individual, can bring into play a power which is not clearly pertaining to him.

The word democracy is derived from the Greek words *demos* and *kratein,* of which the former means the people and the latter means administration or government. This is borne out by earlier descriptions of the conditions sought to be created by democracy.

Status of Education

Aldous Huxley has remarked, "If your aim is liberty and democracy, then you must teach people the arts of being free and of governing themselves." Democracy can never be successful, in the event of lack of education. In a democracy the government is composed of the elected representatives of the people and if the people are uneducated they can never elect the right leaders and consequently can never create the right kind of government. In fact, it is impossible even to hope for democracy in the absence of education. It is difficult to expect a citizen to behave responsibly if he is not even aware of his rights and duties. Bertrand Russell has commented, "Democracy in its modern form would be quite impossible in a nation where many men cannot read." The truth of the matter is that education is a prerequisite of democracy.

Only after proper education should the citizen be invested with his democratic rights. As Fichte, the German philosopher has commented, "Only the nation which has first solved in actual practice the problem of educating perfect men will then solve the problem of the perfect state." Although Fichte made this comment in the context of autocratic states, it cannot be doubted that the perfection of even a democratic state can be judged only by the extent to which it contains educated people.

As Hetherington puts it, "Democratic government, at least, demands an educated people." Throwing light on the objectives of education in the 1949 meeting of the Universities Commission, Dr. Radhakrishnan stressed the fact that the democratic state recognises the importance of the individual, and it is the process of development of this individual which is called education. Hence, education is absolutely necessary for establishing a democratic society. Dewey has pointed out that democracy is inconceivable without education,' because education can generate and instil the qualities which democracy demands as a prerequisite.

Philosophers of the ancient Greek city state were aware of the significance of education. Both Plato and Aristotle laid stress on the importance of education for the success of democracy. Ernest Barker comments, "To Plato education was the most important function of the state and the department of education the most important state department which was particularly advocated for producing the philosopher Kings to improve the men's minds for becoming virtuous beings." Plato, in his famous text, *The Republic,* stressed not only the importance of education for democracy but even formulated a plan for the education of men and women which made all kinds of development -physical, mental, moral and aesthetic-possible.

Aristotle was of the opinion that the aim of the state is to make possible the achievement of the highest moral level and this can be reached through education alone. Thus, education is the most important function of die state. From one point of view, the state itself is a school in which the individual learns citizenship. These truths were known not only to the Ancient Greeks but also to Indian thinkers of ancient times. India has been the home of democratic ideals and principles from ancient times. Hermitages and places of worship were used as schools in which the sages tried to produce ideal citizens who could become useful members of society. But the modem age needs democratic education far more than was needed in ancient Greece or ancient India because modem democracies are so vast and their problems so complex that the education of citizens is even more imperative today.

Different Dimensions

The following are the important aspects or areas of democracy:

Political Democracy: As has already been pointed out earlier, the three ideals of democracy have to be translated into reality in the political, economic and social spheres. It is, therefore, important to understand these three aspects. In its political aspect, democracy can be defined in the words of Lord Bryce, "Democracy is the form of government in which ruling power of a state is legally vested, not in any particular individual or class but in the members of the community as a whole." In a democracy, legally every individual is equal, because law and justice are the same for every one. Although a democratic government is the representative of the majority it takes care to protect every right of the minorities residing within its territories. Every individual has complete right to think as he chooses and to express his opinion. Citizens are not discriminated against on the basis of sex, race, religion or any other ground which distinguishes them from others.

Social Democracy: In its social aspect, democracy lies in emphasis upon equality and brotherhood between all individuals. In the words of Dewey, "Such a society must have a type of education which gives individuals a personal interest in social relationship and control, and habits of mind which secure social change without introducing disorder." The social feeling of a democratic society is best expressed in Kant's famous moral formula, "So act as to treat humanity whether in thine own person or in that of any other, always as an end and never as a means." From this principle Kant derives the practical suggestion, "Try always to perfect thyself and try to conduce to the happiness of others, by bringing about favourable circumstances, as you cannot make others perfect." The democratic individual, therefore, functions as one member of a special government of ends. And this system of ends is the state of democratic ideals. As Kant puts it, "Act as a member of kingdom of ends." Implicit in this is the notion that one should treat oneself and every other individual as something having inherent value, one should act as the member of a state in which every person treats the good of another as no less valuable than his own good. In such a state or country, 911 die others should also behave so that everyone is not a means but an end. Every one should endeavour to promote the good of another while trying to achieve his own ends.

Economic Democracy: The economic aspect of democracy comprehends the rights of individuals to economic independence and equality. In democratic societies all individuals have the right to earn wealth without interfering in or obstructing the rights of other people. This liberty to earn wealth is limited or restricted by the right to equality because, despite the freedom to earn and accumulate wealth, no individual in a democracy has the right to control the means of production in such a way that his possession may obstruct the economic freedom of other people or his wealth may create a chasm between his position and that of others. In order to maintain the right to equality, all democratic societies enact legislation in order to protect the rights of the working class or backward classes. Democracy is opposed to capitalism and it does not approve excessive economic differences between individuals. Because it believes in brotherhood, democracy is opposed to any kind of economic exploitation.

The Basics

According to Sir Cripps, "By democracy we mean a system of government in which every adult citizen is equally free to express his views and desires, upon all subjects in whatever way he wishes and influences the majority of his fellow citizens to decide according to those view and to implement those desires." As far as a definition of a democratic state is concerned, the above definition is quite appropriate. But, as Dr. S. Radhakrishnan pointed out in his University report, democracy is not merely a political system but a way of life which affords equality to every one, irrespective of the differences of race, religion, creed and economic status.

This equality implies equal freedom and equal rights. In fact, democracy is a very comprehensive concept which can be interpreted in three ways -political, social and economic. The creation of democracy depends upon the extent to which the ideals of freedom, equality and brotherhood can be concretised in all three spheres. Before going on to a detailed discussion of democracy in these three spheres, it is necessary to clarify these three fundamental ideals.

Liberty: Liberty is a prerequisite of the success of democracy. John Stuart Mill attached the greatest importance to the liberty of the individual. But liberty does not imply complete wilfulness, for, freedom is not absolute but subject to control imposed by the individual's own conscience. Liberty implies the freedom given to the individual to develop his own abilities as he thinks best, without being conditioned or restricted by any external factor. But the freedom of a large number of people is possible only when no single individual has unlimited power and no individual misuses the rights given to him.

Equality: If democracy is to be successful it is necessary for every individual to be socially, politically and economically equal. Generally speaking, no privileges or special rights should be given to any class of people in a democracy. But equality also does not imply a deliberate neglect of individual differences between one person and another. Stated more accurately, equality actually means an equality of purpose, because equality cannot be used as the basis for ignoring differences between individuals. Lord Haldane was quite correct when he said that one cannot make all people equal because nature is far too powerful. While one woman is born beautiful, another is ugly, and this creates great differences. One man is born with remarkable intelligence, while another is remarkable for the lack of it. Thus the concept of equality should be abandoned, for it has governed the minds of people long enough, with little good effect. In democracy it is perfectly practical to give certain privileges to a backward class of people in order to give them an opportunity to raise themselves to the levels of others, but all such efforts should be exceptions, not the rule. In this way, offices in the administration or government should be open to all those who have the abilities and qualities required for such jobs. Every adult should have the right to vote. Every individual should have the opportunity to get employment which will bring him a wage sufficient to allow him to live well Unemployment, famine and such conditions can prove fatal to democracy.

Fraternity: It is important to achieve the ideal of brotherhood in a democracy because this is the only way of removing the psychological obstacles to democracy. The real difficulty lies not in the establishment of a democratic government but as a means or agency. Brotherhood is the chief characteristic of a democratic society, and cooperation the basis on which members of such a society work together. If a democratic society is to be created it is essential that every effort be made to increase brotherhood and cooperation. G. D. H. Cole is correct when he says, "A democrat is someone who has a physical glow of sympathy and love for any one, who comes to him honestly, looking for sympathy and help; a man is not a democrat, however justly he may try to behave to his fellow men, unless he feels like that." Hence, the establishment of a democratic society requires a complete change in the ideas, personalities and patterns of behaviour of the people. And for this all these people must develop totally. It is difficult even to conceive of democracy in the absence of a liberal mind, sympathy, love, consciousness of social responsibility, a high character and a developed personality.

Social Concerns

Role of Social Education

Education is a lifelong and a never-ending process which continues throughout life. As time passes, changes keep on occurring, and every individual has necessarily to pass through these phases of change. The criterion of success for any individual is the extent to which he can adapt himself to changing circumstances. Keeping in mind this viewpoint, the Education Commission has opined that, in a democracy, the function of adult education is to give to every individual the opportunity for obtaining the kind of education he wants, the kind of education which can contribute effectively to his individual prosperity, professional progress as well as his social and political life. The Education Commission has greatly enlarged the scope of adult education by saying that the field of adult education is as vast as life itself.

India has, for a long time, suffered from a very low percentage of literacy. During the British period, Lord Macaulay's educational policy did much to alleviate this illiteracy, and though, during this period, the British government did, in theory, accept the importance of public education, in practice it did little to achieve this goal. Whatever little was actually achieved was the result of the following three forces:

1. Rise of a middle class in urban areas.
2. The cooperative movement in the villages.
3. Growth of political consciousness among the people.

Of these, it was the third factor which contributed the most, and it is the same political consciousness which has made the future of adult education bright. Till the end of the World War I, satisfactory progress was made in this direction. Adult classes were run exclusively as night classes, and the teacher working in these classes was given an additional allowance. The cooperative movement was started in 1905, and people felt that it was impossible to make it successful without becoming literate. Hence arrangements were made for creating libraries, and stress was laid upon the reading of newspapers. Gradually, universities set up adult education sections in their extension departments.

The growth of adult education can be divided into four periods:

First Phase: During this period, adult education suffered a blow because of the economic depression. The political movement raised the demand for adult franchise. The soldiers in the army were also infused with a strong national spirit when they returned home occasionally from their duties. By 1926, the number of cooperative committees had risen to 80,182. They conducted 100 night schools. In the Punjab, there were 3,784 adult schools in 1926-27, in which 98,414 students were enrolled. In Bombay, there were 27 schools in 1922 while in Uttar Pradesh, 6 municipalities managed night schools for providing adult education. Similar efforts were being made in the other provinces also.

Second Phase: This was a period of political turmoil, but many new schools came into existence during the time. Some missionaries, such as Dr. J.J. Lucas, Dr. J.J.H. Lawrence and Mr. Daniel prepared new books for adults in the Roman script. During this period, not much progress could take place in Punjab, because in 1936-37, it had only 189 schools with an enrolment of 4,988 students. Two experiments, however, were carried out in Punjab.

1. Normal school teachers were inspired to organise libraries in villages for adult education.
2. Cultural centres were set up in villages.

Similar experiments were also carried out in Bombay, Madhya Pradesh, etc., and many new institutions took birth, among them the Rural Reconstruction Association, Adult Education League, Literacy Association, Seva Sadan Social League, etc. The Travancore University began to send sets of books to rural libraries from its own collection for the purpose of adult education.

Bibliography

Adams, D.: *Education and Modernisation in Asia,* Addison Wesley Publishing Company, London, 1970.

Addaval, S.B.: *Theory of Education,* NCERT, New Delhi, 1968.

Aggarwal, J.C.: *Development and Planning of Modern Education with Special Reference to India,* Vikas Publishing House, New Delhi, 1982.

————: *National Policy on Education,* Arya Book Depot, New Delhi, 1979.

Alva, M.: *The Power of Education,* Lancer Books, New York, 1965.

Andrew, W. Halping: *Administrative Theory in Education,* The MacMillan Company, New York, 1967,

Barrie, H.: *Theory and Practice of Education,* Pergamon Press, London, 1968.

Basu, A.: *Education and Political Development in India 1898-1920,* Oxford University Press, Delhi, 1970.

————: *Growth of Education and School Management in India 1898-1920,* Oxford University Press, Delhi, 1970.

Bhatia, S.C.: *Education and Socio-Cultural Disadvantage,* Xerxes Publications, Delhi, 1982.

————: *Education: Theory and Practice Disadvantage,* Xerxes Publications, Delhi, 1982.

Bowman, M.J.: *Education and Economic Development,* Frank Cass, London, 1971.

Carlton, B.: *Foundations of School Management,* Englewood Cliffs, Prentice Hall, New Jersey, 1964.

Cattell, B.B.: *Educational Growth,* Houghton Mifflin Co., Boston, 1971.

Chauhan, S.S. *Principles and Techniques of Education,* Vikas Publishing House, New Delhi, 1982.

Chourasia, G.: *Challenges and Innovations in Education,* Sterling Publishers, New Delhi, 1977.

Cole, S. Brem: *Education and the Development of Nations,* Holt, Rinehart and Winston, New York, 1966.

Crow, A.: *An Introduction to Education Principles and Practices,* Eurasia Publishing House, New Delhi, 1962.

Desai, A.R.: *Social Change and Educational Policy,* NCERT, New Delhi.

Diana, Pinto V.: *Education: Theory, Research and Practice,* Rand McNally College Publishing Company, Chicago, 1978.

Don, Adams: *Education and Modernisation in Asia,* Addison Wesley Publishing Company, London, 1970.

Dube, S.C.: *Educational Planning in India,* Allied Publishers, Bombay, 1965.

————: *Elementary Education in India: A Promise to Keep,* Allied Publishers, Bombay, 1975.

————: *Some Perspectives on Non-Formal Education,* Allied Publishers, New Delhi, 1977.

Eckerson, L.O.: *Guidance Services in Elementary Schools,* Government Printing Press, Washington, 1966.

Ehsanul, Haq: *Education and Political Culture in India,* Sterling Publishers, New Delhi, 1981.

Frank, W.: *Education – Principles and Services,* Charles E. Merrill Books, New York, 1961.

Gandhi, Kishore: *Issues and Choices in Higher Education: A Sociological Analysis,* B.R. Publishing Corporation, Delhi, 1977.

Ghosh, S.C.: *Educational Strategies in Developing Countries,* Sterling Publishers, New Delhi, 1976.

Goel, S.C.: *Education and Economic Growth in India,* The MacMillan Company of India Ltd., Delhi, 1975.

——————: *Education in India,* The MacMillan Company of India Ltd., Delhi, 1975.

Good, W.J.: *World Revolution and Family Patterns,* Collier MacMillan, London, 1963.

Gopinathan, R.: *Education: Population Growth and Socio-Economic Change,* Allied Publishers, New Delhi, 1981.

Gore, M.S.: *Papers in the Sociology of Education in India,* NCERT, New Delhi, 1967.

Gorwaney, N.: *Self Image and Social Change: A Study of Female Students,* Sterling Publishers, Delhi, 1977.

Gunnar, Myrdal: *School Management in India,* Abridged, Allen Lane, London, 1972.

Halping, W.: *Administrative Theory in Education,* MacMillan Company, New York, 1967.

Hanson, J. W. and Cole S. Brem Beck: *Education and the Development of Nations,* Holt, Rinehart and Winston, New York, 1966.

Haragopal, G.: *Administrative Leadership and Rural Development in India,* Light & Life Publishers, New Delhi, 1980.

Henry, W.: *Practice of Education,* McGraw Hill, New York, 1964.

Indu, Prakash Singh: *Women, Law and Social Change in India,* Radiant Pub., New Delhi, 1989.

Iqbal, Narain and Mohan Lal Sarma: *Panchayati Raj and Educational Administration,* Aalakli Publishers, Jaipur, 1976.

Jagannath, Mohanty: *Education for All,* Deep and Deep Pub., New Delhi, 1994.

Jain, Devaki: *Indian Women,* Publication Division, Govt. of India, 1975.

Jones, A.J.: *Principles of Education,* McGraw Hill, New York, 1963.

Joshi, R.N.: *Education – Elsewhere and Here,* Bharatiya Vidya Bhavan, Bombay, 1979.

Kamat, A.R.: *The Educational Situation and other Essays on Education,* People's Publishing House, New Delhi, 1973.

Kidd, J.R.: *Education for Perspective,* Indian Education Association, New Delhi, 1969.

Kiran, Devendra: *Status and Position of Women in India,* Vikas Publishing House, 1985.

Kochhar, S.K.: *Pivotal Issues in Indian Education,* Sterling, New Delhi, 1981.

Kripal, Prem: *A Decade of Education in India,* Indian Book Company, Delhi, 1968.

Lawton, D.: *Education and Social Justice,* Sage Publications, London, 1977.

——————: *Education and School Management,* Sage Publications, London, 1977.

Louis, Malassis: *The Rural World: Education and Development,* Croom Helm, London, 1976.

Minault, Gail: *Secluded Scholars, Women's Education and Muslim Social Reform in Cobaial India,* Oxford University Press, Delhi, 1998.

Misra, R.S.: *Women Education and the Upanishadic System of Education,* Chugh Publication, Allahabad, 1993.

Moni, Mohan, Bose: *Female Education in India,* B. B. Gupta Publication, Kanpur, 1921.

Mukherji, S.N.: *Administration of Education in India,* Acharya Book Depot, Baroda, 1962.

Narayan, G. Brij Raj Chauhan and T.R. Singh: *Scheduled Castes and Education,* Anu Publications, Meerat, 1975.

Niblett, W.R.: *Essential Education,* University of London, London, 1955.

Nural, H. S.: *Challenges in Education: Culture and Social Welfare,* Allied Publishers, Bombay, 1977.

Oldhan, J.N.: *Village Education in India,* Oxford University Press, London, 2000.

Pande K.C. and Mohan Lal Sarma: *Panchayati Raj and Educational Administration,* Aalakli Publishers, Jaipur, 1976.

Paulo, Freire: *Education: The Practice of Freedom,* Writers and Readers Publishing Co-operative, London, 1976.

Prakash, Shri: *Educational System of India: An Econometric Study,* Concept Publishing Company, Delhi, 1977.

Prem Kripal: *A Decade of Education in India,* Indian Book Company, Delhi, 1968.

Premi, M.K.: *Educational Planning in India,* Sterling Publishers, New Delhi, 1972.

Rajagopal, M.V.: *Kothari Commission on School Education,* Telugu Vidyardhi Prachuranalu, Machilipatnam, 1967.

Ram, Mohan Das: *Women in Manu's Philosophy,* ASB Pub., Jalandhar, 1993.

Ram, S. Sharma: *Women's Education in Ancient and Muslim Period,* Discovery Publishing House, New Delhi, 1996.

Raman, Lal Mehta: *Legal Rights of Women under Different Communal Laws in Vogue in India,* The Advocate of Indian Press, Bombay, 1933.

Rao, M.S.: *Sociology of Education in India,* NCERT, New Delhi, 1977.

Rao, V.K.R.V.: *Education and Human Resource Development,* Allied Publishers, Bombay, 1966.

Rehana, Ghadially: *Women in Indian Society,* Sage Pub., New Delhi, 1988.

Robert, H.: *Educational Policy and Practice,* Harper and Row, New York, 1962.

Rudolph and Rudolph: *Administration and Management in Schools,* Harvard University Press, Cambridge, 1972.

————————: *Education and Politics in India Studies in Organisation, Society and Policy,* Harvard University Press, Cambridge, 1972.

Russell, Bertrand: *Education and the Social Order,* Unwin, London, 1977.

Safaya, Raghunath: *Innovations and Latest Trends in Education,* The Associated Publishers, Ambala Cantt., 1976.

————————: *Latest Trends in Education,* The Associated Publishers, Ambala Cantt., 1976.

Saini, S.K.: *Development of Education in India: Socio-Economic and Political Perspective,* Cosmo Publications, New Delhi, 1980.

Saran, Gunam: *Education and Social Change: A Study of Some Rural Communities in India,* The Minarva Associates, Calcutta, 1972.

Sateswari, Saxena: *Education Planning in India: A Study in Approach and Methodology,* Sterling Publishers, New Delhi, 1974.

Shah, A.B.: *Education or Catastrophe?* Vikas Publishing House, New Delhi, 1976.

Sharma, G.S.: *Educational and Cultural Development of the Minorities,* ICSSR, New Delhi, 1973.

______________: *Educational and Cultural Development of the Minorities – A Study in Social Effects of the Judicial Trends in India,* ICSSR, New Delhi, 1973.

Shipman, M.D.: *Education and Modernisation,* Faber and Faber London, 1971.

Shrimali, K.L.: *A Search for Values in Indian Education,* Vikas Publishers, Delhi, 1974.

Shukla, P.D.: *Towards the New Pattern of Education in India,* Sterling Publishers, New Delhi,1976.

Singhal, R.P. and Biswas Dutt Sunnittee: *The New Educational Pattern in India,* Vikas Publishing House, Delhi, 1916.

______________: *The New Educational Pattern in India,* Vikas Publishing House, Delhi, 1916.

Singla, M.M.: *Management of School,* The World Press, Calcutta, 1975.

Sinha, P.: *Descriptive-cum-Critical Study of Education System in Andhra Pradesh,* Administrative Staff College of India, Hyderabad, 1976.

Siqueira, T.N.: *The Education of India: History and Problems,* Oxford University Press, London, 1952.

Smith, H M. and Eckerson, L.O.: *Guidance Services in Elementary Schools,* Government Printing Press, Washington, 1966.

Sundaram, P.S. and A.B. Shah: *Education or Catastrophe?* Vikas Publishing House, New Delhi, 1976.

Sundaram, P.S.: *Education: Theory and Practice,* Vikas Publishing House, New Delhi, 1976.

Swamy, L. and Mudaliar A.: *Education in India,* Asia Publishing House, Bombay, 1960.

Tandon, R.K.: *Women, Nature, Education, Teaching and Rights,* Commonwealth Pub., 1996.

Thomas, F.: *School Management and Administration,* MacMillan Co., New York, 1953.

Tiwari, D.D.: *Education at the Cross Roads,* Chugh Publications, Allahabad, 1975.

—————: *Thoughts on Education,* Chugh Publications, Allahabad, 1972.

Tolman, E.C.: *Principles of Purposive Behavior,* Ajanta Publication, Delhi, 1988.

Ulrich, B.: *Development and Planning of Modern Education,* Vikas Publishing House, New Delhi, 1982.

Uttam, Kumar and Nayak A.R.: *Women Education,* Commonwealth Pub., New Delhi, 1996.

Vaizey, John: *Education for Tomorrow,* Penguin, London, 1966.

Vernon, P.E.: *School Administration in India,* Methuen, London, 1969.

William, Evan M.: *Organizational Theory Structure, Systems and Environments,* Wiley Interscience Publications, John Wiley and Sons, London, 1976.

Yvonne, Basrin: *Woman of Tomorrow,* Harvard University Press, New York, 1985.

Zurich, D.: *Development of Education,* Allied Publishers, Bombay, 1978.

Sundaram, P.S.: *Education: Theory and Practice*, Vikas Publishing House Pvt., Delhi, 1976.

[illegible] and Mutalik, [illegible]: *Education in* [illegible], [illegible] Publishing House, Bombay, [illegible].

[illegible], K.C.: [illegible] *Education: Theory and* [illegible], [illegible], Delhi, 1978.

Thomas, [illegible]: *School Management and Administration*, [illegible] & Co., [illegible]

Tiwari, D.D.: *Education* [illegible], Chugh Publications, Allahabad, 1972.

[illegible]: [illegible], Chugh Publications, Allahabad, 1972.

[illegible], R.C.: [illegible]

[illegible]

[illegible]

[illegible]

[illegible], London, 1969.

[illegible], 1976.

[illegible] University Press, [illegible]

[illegible] Allied Publishers, [illegible]

Index

A

Abraham Lincoln, 281.
Adult Education, 2, 3, 4, 5, 9, 10, 11, 12, 13, 14, 15, 16, 17, 18, 21, 22, 23, 24, 25, 26, 27, 28, 29, 31, 32, 33, 34, 35, 36, 198, 287, 288.
Association, 31, 68, 69, 78, 84, 117, 138, 246, 255, 288.
Australia, 19, 95.

B

B.R. Ambedkar, 125.
British, 87, 128, 140, 177, 178, 181, 182, 203, 210, 214, 287.
Broadcasting, 81.
Buddha, 136, 141, 144.

C

California, 9.
Canada, 88.
Central Education Advisory Council, 10, 31.
Central Pollution Control Board, 266.
Chancellor, 215.
Chandra Mohanty, 156.
Constitutional Right, 281.

D

Democracy, 28, 29, 33, 111, 113, 179, 183, 206, 224, 234, 235, 236, 281, 282, 283, 284, 285, 286, 287.
Dharma, 73, 130, 141, 142, 144, 145.
Dowry Prohibition Act, 191, 216.

E

Editor, 9.
Education Commission, 1, 6, 12, 13, 18, 19, 21, 64, 69, 101, 287.
England, 122, 211, 224.
Environmental Pollution, 265.
Europe, 89, 137.
Exercise, 40, 86, 94, 108, 110, 114, 121, 122.

F

France, 87, 224, 244.
Freedom, 81, 92, 104, 115, 122, 123, 154, 178, 201, 204, 209, 212, 213, 214, 235, 247, 257, 270, 274, 282, 285, 286.

G

Gandhi, 145, 212, 213, 222.
Germany, 244.
Gita, 135, 144, 145.
Guidelines, 106, 113, 118, 199.

H

Hindu, 72, 73, 75, 94, 120, 134, 143, 144, 145, 180, 210, 211, 216, 253, 277.
Hostility, 40, 151.

I

Illiteracy, 5, 6, 10, 12, 14, 15, 17, 21, 29, 31, 32, 33, 34, 67, 159, 161, 168, 214, 287.
Indian Thinkers, 140, 283.
Interview, 217.

J

John Locke, 150.
Journalism, 202.

K

K.G. Sayyidain, 26.
Karma, 141, 144, 145.

L

Liberty, 235, 239, 281, 282, 285, 286.
Life Insurance Corporation, 259.
Linda J. Nicholson, 153.
Literacy, 1, 2, 3, 5, 6, 7, 10, 11, 12, 13, 14, 15, 16, 17, 20, 22, 23, 24, 25, 26, 28, 30, 31, 32, 33, 34, 35, 36, 62, 168, 169, 188, 189, 196, 225, 230, 233, 287, 288.
Literature, 9, 10, 18, 22, 24, 30, 32, 33, 100, 102, 150, 215, 216.

M

Mahabharata, 73, 125, 144.
Media, 17, 29, 77, 90, 98, 167, 181, 190, 202, 206, 207, 229.

N

Nancy Chodorow, 152.
National Capitai Region, 265.
National Education Commission, 1.
News, 81.
Newspapers, 9, 64, 81, 99, 244, 248, 258, 288.

O

Organisation, 4, 8, 10, 12, 13, 15, 29, 38, 42, 47, 48, 50, 53, 54,

55, 64, 69, 109, 123, 129, 130, 132, 137, 140, 172, 186, 189, 190, 207, 219, 221, 245, 255, 282.

P

Parliament, 139, 180, 215, 241, 271.

Press, 279.

Publications, 9.

R

Ramayana, 125, 144.

Rural Community, 251.

Russia, 16, 137, 215, 224.

S

Social Education, 1, 2, 3, 4, 5, 6, 7, 8, 9, 10, 24, 25, 26, 27, 28, 29, 30, 31, 32, 33, 34, 164, 226, 287.

Social Equality, 71, 281.

Social Welfare, 7, 15, 16, 157, 158, 164, 171, 172, 173, 174, 184, 189, 198, 202, 203, 205, 217, 224.

Speech, 248.

Supreme Court, 119, 265.

T

Tamil Nadu, 229, 252, 255, 266.

Television, 3, 6, 7, 17, 19, 20, 258.

Text, 283.

Transmission, 141, 235.

U

UNESCO, 21, 29, 30.

United States, 60, 95, 103, 106, 215, 266.

University, 9, 11, 12, 20, 21, 30, 77, 116, 117, 173, 215, 257, 285, 288.

USSR, 7, 19, 80, 81, 90.

V

Varahamihira, 120.

Vedic Period, 145, 216.

W

Williams, 249, 272.

World Bank, 231.

Y

Yajnavalkya, 73, 144.

Yashwant Doshi, 9.

Z

Zone, 266.

❑❑❑